GAS SEPARATION

TECHNIQUES, APPLICATIONS AND EFFECTS

CHEMISTRY RESEARCH
AND APPLICATIONS

Additional books and e-books in this series can be found on Nova's
website under the Series tab.

CHEMISTRY RESEARCH AND APPLICATIONS

GAS SEPARATION

TECHNIQUES, APPLICATIONS AND EFFECTS

SURAYA MATHEWS
EDITOR

nova
science publishers
New York

NOTICE TO THE READER

Additional color graphics may be available in the e-book version of this book.

Library of Congress Cataloging-in-Publication Data

Names: Mathews, Suraya, editor.
Title: Gas separation : techniques, applications, and effects / editor, Suraya Mathews.
Description: Hauppauge, New York : Nova Science Publishers, Inc., 2018. | Series: Chemistry research and applications | Includes bibliographical references and index.
Identifiers: LCCN 2018047553 (print) | LCCN 2018048613 (ebook) | ISBN 9781536146073 (ebook) | ISBN 9781536146066 (softcover)
Subjects: LCSH: Gases--Separation.
Classification: LCC TP242 (ebook) | LCC TP242 .G37285 2018 (print) | DDC 660/.043--dc23
LC record available at https://lccn.loc.gov/2018047553

Published by Nova Science Publishers, Inc. † New York

CONTENTS

PREFACE

In recent decades, the science of gas separation by use of a nanoporous permselective membrane has widely developed due to properties such as low energy consumption, easy operation, low waste generation and economic benefits. In Gas Separation: Techniques, Applications and Effects, the fundamental concepts of membrane gas separation and the formation of nanoporous membranes are been discussed.

The authors go on to examine mixed matrix membrane, a composite material comprised of inorganic fillers. The primary role of fillers is to systematically manipulate the molecular packing of the organic phase, thus enhancing the gas separation properties of matrix membranes.

The closing study analyzes the permeability and selectivity of carbon dioxide and methane gas of polyvinylchloride mixed matrix membrane with the inorganic fillers of zeolite 4Å particles. The fabrication of mixed matrix membranes is prepared by using dry/wet phase inversion method, and Fourier transform infrared spectroscopy is used to study the chemical interaction of the membrane by analyzing the intensity of the peak of chloride vibration.

Chapter 1 - In recent decades, the science of gas separation especially by using a nanoporous perm selective membrane has widely developed due to the peculiar properties such as low energy consumption, easy operation, low waste generation and economic benefits. The most important membrane gas separation applications include H_2 production from syngas, offshore gas-processing platforms, separating atmospheric gas for medical and industrial utilization and isotope separation for nuclear power uses. In order to achieve high efficiency in the gas separation process, a broad range of materials was investigated to explore a membrane that exhibits high selectivity and permeability to various gases. This chapter allocated to the fundamental concepts of membrane gas separation such as various mechanisms of gas separation and subsequently the formation of nanoporous membranes and different types of membranes and their structure have been discussed. The final section of this book chapter contained the basic concepts and processes of modeling and simulation for various gas separation utilizing nanoporous membranes.

Chapter 2 - Mixed matrix membrane (MMM) is a composite material comprising organic phase and inorganic fillers. The primary role of fillers is to systematically manipulate the molecular packing of the organic phase, thus enhance the gas separation properties of MMMs. For the past few decades, the emphasis was mainly placed on the incorporation of inorganic fillers such as zeolite, carbon molecular sieve, and silica. These fillers are commonly used due to their narrow pore size which can yield high selectivity. Nevertheless, it is often difficult to produce MMM with defect-free morphologies using these fillers due their poor distribution properties and interfacial adhesion with the polymer matrix, which deteriorate their gas separation performance. Consequently, research has expanded to alternative fillers such as layered silicate, carbon nanotube (CNT), graphene oxide (GO) polyhedral oligomeric silsesquioxane (POSS) and titanium dioxide (TiO_2). These alternative fillers are predominantly used in the preparation of nanocomposites with increased thermal, chemical and mechanical stability. These enhancements have improved the engineering capability of nanocomposites to be used in a wide range of application which includes elastomers, thermoset plastics, thermoplastics, materials for drug delivery, dental and medical applications. However, the use of alternative fillers for gas separation MMM synthesis is still limited. Carbonaceous nanofiller with organized carbon structure has been found to exhibit outstanding reinforcing properties on the polymer matrix. Despite its unique morphological and structural properties, limited research has been conducted using carbonaceous nanofiller for MMM synthesis. Thus far, the application of carbonaceous nanofillers such as carbon nanotubes (CNT) and graphene oxide (GO) has shown promising result in terms of gas separation performance. However, the separation efficiency of CNT-based MMMs can only be achieved if CNTs are aligned vertically. Likewise, the promising result by GO is challenged by the bulk quantity production of the graphene sheet. Recently, polyhedral oligomeric silsesquioxane (POSS) is receiving immense attention due to its monodispersed size, low density, and ease of modification. The truly unique feature of POSS lies in its inner inorganic and outer organic framework, which can be made up of various organic or inorganic substituents. The hybrid properties it possesses allows POSS to be dispersed homogeneously in the polymer matrix without agglomeration or aggregation. Hence, accompanied by the intrinsic molecular property of POSS, the performance of the POSS-based MMMs can be enhanced. Nonetheless, the performance of these MMMs is dictated by the incorporation technique employed during membrane fabrication. TiO_2 is an emerging non-porous metal oxide filler for membrane separation application. Its inherent advantage comes from its ability to be dispersed individually. The addition of optimum loading of metal oxides may lead to significant enhancement in the permeability of the membrane while maintaining or enhancing membrane selectivity. On the other hand, the overall permeability and selectivity may also decrease due to the non-porous nature of metal oxides which are unable to selectively sieve out molecules of different sizes. Much recent work has gone into attempting to tailor novel alternative fillers which could be incorporated into MMM for gas separation application.

Chapter 3 - Mixed-Matrix Membrane is a developing technology that has been use in the gas separation process due to the ability of MMMs to cope with the limitation of polymeric membrane and inorganic membrane. Therefore, this research is conducting to study the permeability and selectivity of carbon dioxide (CO_2) and methane gas (CH_4) of polyvinylchloride (PVC) Mixed-Matrix-Membrane (MMMs) with the inorganic fillers of zeolite 4Å particles. The fabrication of MMMs is prepared by using dry/wet phase inversion method. Fourier Transform Infrared Spectroscopy (FTIR) is used to study the chemical interaction of the membrane by analyzing the intensity of the peak of chloride vibration. Meanwhile, Scanning Electron Microscope (SEM) is use to analyses the cross sectional morphology of MMMs. The performance of MMMs analyses by using Design of Expert (DOE) method. While, the model regression equation is developed as the potential use for screening the permeability of CO_2 and CH_4 based on the effect of PVC and zeolite concentration.

In: Gas Separation
Editor: Suraya Mathews

ISBN: 978-1-53614-606-6
© 2019 Nova Science Publishers, Inc.

Chapter 1

MODELING AND SIMULATION OF NANOPOROUS MEMBRANES FOR GAS SEPARATION

Atyeh Rahmanzadeh and Masoud Darvish Ganji[*]
Department of Nanochemistry,
Faculty of Pharmaceutical Chemistry,
Tehran Medical Sciences, Islamic Azad University,
Tehran, Iran

ABSTRACT

In recent decades, the science of gas separation especially by using a nanoporous perm selective membrane has widely developed due to the peculiar properties such as low energy consumption, easy operation, low waste generation and economic benefits. The most important membrane gas separation applications include H_2 production from syngas, offshore gas-processing platforms, separating atmospheric gas for medical and industrial utilization and isotope separation for nuclear power uses. In order to achieve high efficiency in the gas separation process, a broad range

[*] Corresponding Author E-mail: ganji_md@yahoo.com.

of materials was investigated to explore a membrane that exhibits high selectivity and permeability to various gases. This chapter allocated to the fundamental concepts of membrane gas separation such as various mechanisms of gas separation and subsequently the formation of nanoporous membranes and different types of membranes and their structure have been discussed. The final section of this book chapter contained the basic concepts and processes of modeling and simulation for various gas separation utilizing nanoporous membranes.

1. INTRODUCTION

Industrial gas separation depends on three main processes - pressure swing adsorption (PSA), cryogenic distillation and membrane gas separation. PSA technology is utilized to separate a target gas species from a mixture of gases under pressure, according to the species' molecular properties and affinity for an adsorbent material. Particular adsorptive materials like zeolites are used as molecular sieves, preferentially adsorbing the target gas species at high pressure. Then, this process swings to low pressure to desorb the adsorbent material. The PSA process is the most widely used industrial process to separate H_2 from a mixture of gases [1]. However, industrial applications often need PSA gas purification processes to be operated at high temperatures, which decreases the sorbent selectivity for the desired species and reduces the performance of the gas purification process. Thus, significantly lower flow rates are used, in order to achieve a high performance [2].

Cryogenic distillation process separates the target gas species from a mixture of gases at low temperatures based on the diversity in boiling temperatures of the gas components [3]. As respects, cryogenic distillation requires a phase variation, and thus consumes a significant amount of energy [1, 4, 5].

Other than PSA and cryogenic separation, membrane separation techniques have enticed the greatest interest. Membranes are obstacles that only permit selected materials to permeate across them. Membrane separation processes offer multifold advantages over the more mature and commercially-available PSA and cryogenic separation processes [1].

1.1. Brief History of Gas Separation Membranes

Long before the first commercial gas separation membranes (named Prism) were introduced, people had already noticed the potential usage of membrane as gas separation tools. In 1829, Thomas Graham discovered the law of gas diffusion by utilizing a tube with one end sealed with plaster of Paris [6]. Three years later, Mitchell [7] reported for the first time that different gas molecules have various tendency to pass through rubber membranes, which means the flux of each gas is different. Since then, lots of polymers have been studied extensively to look for their potential to be gas separation membranes. H. A. Dynes and R. M. Barrer are the pioneers in performing quantitative measurements of gas permeability by using the time-lag method. A number of permeability data had been obtained from lots of potential membrane material [8]. However, the lack of technology to produce high performance and low cost modules postponed the application of gas separation membranes, until Loeb and Sourirajan [9] invented a new phase inversion method to cast asymmetric cellulose acetate membranes, which enables the reduction of effective membrane thickness from several micrometers into sub-micrometer level. The invention of high-flux anisotropic membrane modules in the forms of spiral-wound and hollow fiber further facilitated the development of gas separation membranes. In 1980, Permea delivered the first generation polysulfone hollow-fiber membranes for hydrogen recovery from purge gas steams of ammonia plants. Soon after the success of Permea, Cynara (now part of Natco), Separex (now part of UOP), and GMS (now part of Kvaerner) had commercialized cellulose acetate membranes for removing carbon dioxide from natural gas [10]. More recently, the Polaris^{TM} membrane developed by MTR (Membrane Technology and Research, Inc.), which is a thin film composite membrane in spiral wound form, shows a CO_2 permeance ten times higher than the conventional cellulose membranes [11]. Many fundamental scientific works and contributions related to gas separation membranes were carried out in the twentieth century, as summarized in Table 1 [12].

Table 1. Scientific developments of membrane gas transport [7]

Scientist (Year)	Event
Graham (1829)	First recorded observation
Mitchell (1831)	Gas permeation through natural rubbers
Fick (1855)	Law of mass diffusion
von Wroblewski (1879)	Permeability coefficient product of diffusion and absorption coefficient
Kayser (1891)	Demonstration of validity of Henry's Law for the absorption of carbon dioxide in rubber
Lord Rayleigh (1900)	Determination of relative permeabilities of oxygen, nitrogen and argon in rubber
Knudsen (1908)	Knudsen diffusion defined
Shakepear (1917-1920)	Temperature dependence of gas permeability independent of partial pressure difference across membranes
Daynes (1920)	Developed time lag method to determine diffusion and solubility coefficient
Barrer (1939-1943)	Permeabilities and diffusivities followed Arrhenius equation
Matthes (1944)	Combined Langmuir and Henry's law sorption for water in cellulose
Meares (1954)	Observed break in Arrhenius plots at glass transition temperature and speculated about two modes of solution in glassy polymers
Barrer, Barrie and Slater (1958)	Independently arrived at dual mode concept from sorption of hydrocarbon vapors in glassy ethyl cellulose
Michaels, Vieth and Barrie (1963)	Demonstrated and quantified dual mode sorption concept
Vieth and Sladek (1965)	Model for diffusion in glassy polymers
Paul (1969)	Effect of dual mode sorption on time- lag and permeability
Petropoulos(1970)	Proposed partial immobilization of sorption
Paul and Koros (1976)	Defined effect of partial immobilizing sorption on permeability and diffusion time lag

As a consequence, the successful application of the first commercial gas separation membrane has accelerated the development of novel membrane materials as it offer an attractive alternative for specific separation applications.

1.2. Basic Concept of Membrane Separation

The description of a membrane can be explained as a semipermeable active or passive barrier which, under a definite driving force, permits preferential passage of one or more species or components (molecules, particles or polymers) in a gaseous and/or liquid mixture solution [13, 14].

The initial species rejected by the membrane is named retentate or sometimes just ''solute", while those species passing through the membrane is usually termed "permeate" or sometimes "solvent" shown in Figure 2. The driving force can stabd in the form of pressure, concentration, or voltage difference through the membrane [13]. Depending on the driving force and the physical sizes of the separated species, membrane processes are categorized accordingly: microfiltration (MF), ultrafiltration (UF), nanofiltration (NF), reverse osmosis (RO), dialysis, electrodialysis (ED), pervaporation (PV), and gas separation. Table 2 shows the size of materials retained for each process, the driving force behind separation and the type of membrane, while Table 3 summarizes some of the applications of each process and their alternatives [14,15].

Membrane processes can be operated in two main modes according to the direction of the feed stream relative to the orientation of the membrane surface: dead-end filtration and cross-flow filtration (see Figure 2). The majority of the membrane separation applications utilize the concept of cross flow where the feed flows parallel to and past the membrane surface while the permeate penetrates through the membrane overall in a direction normal to the membrane. The shear force applied by the flowing feed stream on the membrane surface help to remove any stagnant and accumulated rejected species that may reduce permeation rate and increase the retentate concentration in the permeate. Predominant in the regular filtration process, dead-end filtration is used in membrane separation only in a few cases such as laboratory batch separation. In this mode, the flows of the feed stream and the permeate are both perpendicular to the membrane surface [17].

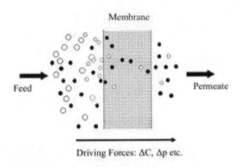

Figure 1. Schematic diagram of gas separation process by a membrane.

Table 2. Size of materials retained, driving force and type of membrane used for each separation process [15]

Process	Size of retained material	Driving force	Type of membrane
Microfilteration	0.1-10μm	Pressure difference (0.5 - 2 bar)	Porous
Ultrafiltration	1-100nm	Pressure difference (1 - 10 bar)	Microporous
Nanofiltration	0.5-5nm	Pressure difference (10 - 70 bar)	Microporous
Reverse Osmosis	<1nm	Pressure difference (10-100 bar)	Non-porous
Dialysis	<1nm	Concentration difference	Non-porous or microporous
Electrodialysis	<1nm	Electrical potential difference	Non-porous or microporous
Pervaporation	<1nm	Concentration difference	Non-porous
Gas Permeation	<1nm	Partial pressure difference (1-100 bar)	Non-porous

Table 3. Examples of membrane applications and alternative separation processes [16]

Process	Applications	Alternative Processes
Microfiltration	Separation of bacteria and cells from solutions	Sedimentation, Centrifugation
Ultrafiltration	Separation of proteins and virus, concentration of oil-in-water emulsions	Centrifugation
Nanofiltration	Separation of dye and sugar, water softening	Distillation, Evaporation
Reverse Osmosis	Desalination of sea and brackish water, process water purification	Distillation, Evaporation, Dialysis
Dialysis	Purification of blood (artificial kidney)	Reverse osmosis
Pervaporation	Dehydration of ethanol and organic solvents	Distillation
Gas Permeation	Hydrogen recovery from process gas streams, dehydration and separation of air	Absorption, Adsorption, Condensation
Membrane Distillation	Water purification and desalination	Distillation

Figure 2. Schematic diagram of (left) dead-end filtration and (right) cross-flow Filtration.

Two of the most essential parameters that explain the separation performance of a membrane are its permeability and permselectivity (or simply as selectivity). Permeability is usually used to support an indication of the capacity of a membrane for processing the permeate; a high permeability means a high throughput. A high throughput is useless, however, unless another membrane feature, permselectivity also exceeds an economically acceptable level. On the other hand, a membrane with a high permselectivity but a low flux or permeability may require such a large membrane surface area that it becomes economically unattractive. Simply put, permselectivity is the ability of the membrane to separate the permeate from the retentate.

As a general rule, the membrane technology is a competitive separation method for small to medium volumetric flow rate functions and for either primary separation or separation with a requirement of purity level of 95%~99% [13, 15, 16].

1.3. Mechanisms

Membrane gas separation is a pressure-driven process that can be attributed to four mechanisms: (i) Knudsen diffusion, (ii) molecular sieving, (iii) solution - diffusion and (iv) surface diffusion [1, 13, 18].

1.3.1. Knudsen Diffusion

Knudsen diffusion is considered in porous membranes, whose pore size are less than mean free path of the gas molecule [13]. Gas molecules therefore interact with the pore walls much more frequently than with one another and allow lighter molecules to preferentially diffuse through pores to achieve separation. Knudsen diffusion principally takes place in the membranes with the pore size of 50-100Å in diameter [19].

Knudsen diffusion occurs when the permeating species flow *via* the membrane almost independent of one another. Hence, the Knudsen diffusion coefficient, D_k (m²/s) is independent of pressure. For an equimolar feed, the permeation rate of Knudsen diffusion is inversely proportional to the square root of the molecular weight of the various compounds in the following equation [20]:

$$D_k = 0.667rv = 97r\sqrt{\frac{T}{M_w}} \tag{1}$$

where the average pore radius is given by r (m), v is the average molecular velocity (m/s) and T is the operating temperature. Whilst, the highest attainable separation factor between two different gas molecules i and j equals to the square root of the ratio of the two gas molecular weights, shown in equation (2).

$$\alpha_{ij} = \sqrt{\frac{M_{wj}}{M_{wi}}} \tag{2}$$

As a result, such membranes are not commercially attractive in general for standard application due to their relative low selectivity.

1.3.2. Molecular Sieving

Molecular sieving separation is first and foremost based on the precise size discrimination between gas molecules through ultramicropores (< 7Å in diameter). Molecular sieving membranes become increasingly important in gas separation especially for inorganic membranes due to their higher productivity and selectivity than solution-diffusion polymeric membranes [21, 22]. Their porous nature has led to high permeability, while the high selectivity is achieved through effective size and shape separation between the gas species. This happens when the pore diameters are small enough to allow the permeation of smaller molecules while obstructing the larger molecules to diffuse through. Even both can enter the pores; the larger one would experience stronger repulsive forces. This energetically biased selectivity is called "energetic selectivity." It is also believed that the more subtle contribution to selectivity is from "entropic selectivity" [23], in which the rotational freedom of one component is restricted to a higher degree than that of the other components. Carbon molecular sieve membranes (CMSMs) and zeolites are the typically membranes dominated by molecular sieving mechanism and give the high separation performance. The ratio of the gas molecular size to the micropore diameter controls the gas permeation rate and separation in molecular sieving materials [24]. For example, zeolite 4A with a pore size of 3.8Å has an O2 and N2 selectivity of approximately 37 at 35°C [25]. The CO2/CH4 ideal selectivity has reached as high as around 200 in the carbon molecular sieve membranes [26]. The pore size of zeolite may be controlled or modified by choosing suitable synthesis, dealumination, and ionexchange [27]; as for CMSMs, the pore size can be controlled by choosing polymer precursor, pyrolysis conditions, pretreatment and post treatment [28].

1.3.3. Solution –Diffusion

The solution- diffusion mechanism, through the selective layer of a nanoporous membrane occurs in the absence of direct continuous pathways for the transportation of gas penetrants across the membrane. This transport mechanism produces high performance membranes, which are used exclusively in commercial separation devices to conveniently separate wide spectrum of gas pairs. Solution-diffusion mechanism is conceptually assumed that gas molecules from the upstream gas phase first adsorb into the membrane, then diffuse across it and finally desorb into the downstream gas phase side [14, 23]. This mechanism commonly found in the gas transportation through polymeric membranes.

The permeation of molecules through membranes is verified by two main mechanisms. They are diffusivity (D) and solubility (S) [24]. Diffusivity is the mobility of lone molecules passing through the holes in a membrane material, and solubility is the number of molecules dissolved in a membrane material. Permeability (P) is defined as an ability measurement of a membrane to permeate molecules:

$$P = D \times S \tag{3}$$

The ability of a membrane to separate two molecules, A to B, is the ratio of their permeability, called as the membrane selectivity αAB.

$$\alpha_{AB} = \frac{P_A}{P_B} = \frac{D_A S_A}{D_B S_B} \tag{4}$$

1.3.4. Surface Diffusion

The diffusion coefficients of common gases in polymers were recognized early as a strong application of the effective diameter of the gas molecule. DA/DB is the ratio of the diffusion coefficients of the two molecules, and can be interpreted as the mobility or diffusivity selectivity, reflecting the different sizes of the two penetrating molecules.

Diffusion in rubbery polymeric materials involves the generation of a sufficiently large gap in the polymer adjacent to the sorbed penetrant for the penetrant to move into, with the subsequent collapse of the sorbed cage that was previously occupied by the penetrant. Thermally induced motions of the polymer segments are responsible for creation and destruction of these transient gaps or holes. The rate of diffusion depends on the concentration of holes that are sufficiently large to accept diffusing molecules.

Diffusion in glassy polymeric materials is different from rubbery polymers primarily because of the difference in the characteristic scales of the micromotions that occur at 30 segmental level for the two states. In glassy polymers the micromotions are much less extensive than rubbery polymers and are believed to be torsional oscillations. Some differences arise due to the presence of trapped excess free volume in glassy materials.

It has been well established that the diffusion coefficient is the initial factor in determining the precise value of gas permeability in polymers [25]. The diffusivity of gases was shown to decrease quickly as the collision diameter of the gas molecule (determined from gas viscosity data) increases. According to the study by Berens and Hopfenberg [26], the diffusivities of a wide variety of gases and vapors in poly (vinyl chloride) exhibited a systematic progression. The diffusion coefficient changed ten orders of magnitude with an order of magnitude change in diameter. Other molecular size parameters suggested include molar volume, square root of molecular weight, and kinetic or *Lennard-Jones* (L-J) diameter. The interactional relationship of these quantities may give different results. A specific example is CO_2 which has a low kinetic diameter but a larger molar volume or molecular weight square root [27].

2. NANOPOROUS MEMBRANES

2.1. Membrane Formation

Dense homogeneous membranes are frequently used in laboratory research to characterize the intrinsic permeation properties. They are normally prepared by the solvent casting or melt extrusion techniques. For the solvent casting technique, the polymer solution with a certain viscosity is cast on a flat plate followed by solvent evaporation at a given temperature. For polymers such as polyethylene, polypropylene and polyamide that are difficult to dissolve in solvents, the membranes can be produced by the melt extrusion technique. The membranes are formed by compressing the polymers between two heated plates at a temperature just below the melt point of the polymers [28]. Most of the membranes for gas separations in industry are asymmetric or composite membranes [13]. These membranes have a very thin selective layer, formed by solvent casting or dip coating, supported on a porous substrate in order to achieve a high permeation flux. According to the solution-diffusion mechanism and the definition of the permeance (Eq. 5), a thin membrane effective thickness favors the permeation rate, as well as the productivity of the whole process. In this equation, the permeance J is given as:

$$J = DK_D \frac{\Delta P}{L} \tag{5}$$

where L is the effective membrane thickness, ΔP is the pressure difference between upstream and downstream sides, D is the diffusion coefficient and K_D is the sorption coefficient.

2.1.1. Asymmetric Membranes

Asymmetric membranes (Figure 3) as layered structures in which the porosity, pore size or the membrane composition changes gradually from one side to the other side of the membrane. The membranes are normally prepared by a phase inversion process, in which a polymer solution is separated into two phases: a solid, polymer-rich phase that forms the matrix of the membrane, and a liquid, polymer-poor phase that forms the pore of the membrane [29].

Figure 3. Structure of membranes used in membrane-based gas separation .
Left: Asymmetric membrane, right: Composite membrane [30].

2.1.2. Composite Membranes

Composite membranes (Figure 3) are formed primarily for two reasons: to seal defects on the surface of an asymmetric membrane ("resistance model" composite membrane), or to form a dense selective layer on the top of a porous substrate (thin-film composite membrane). Several methods have been developed to prepare composite membranes: solution coating, interfacial polymerization, thin film lamination and plasma polymerization [31, 32].

The membrane formation is not the key issue of discussion within the frame of the thesis study. For sake of convenience, the membranes used in simulation of this thesis are all assumed to be homogeneous in thickness, in other words, the effective membrane thickness is considered as equal to the real fiber's thickness and the permeation parameters are independent on the membrane thickness.

2.2. Membrane Modules for Gas Separation

The core part of any membrane is the module, *i.e.,* the technical arrangement of membrane. Three major types of modules can be distinguished in Figure 4.

Figure 4. Schematic membrane modules. Left: Plate and frame module. Middle: Spiral-wound module. Right: Hollow fiber module (http://www.co2crc.com.au).

2.2.1. Plate and Frame

The plate and frame modules are one of the earliest types of membrane modules. Simply, a planar membrane separates the permeate and the retantate flows. It results in relatively high cost and low packing density. Nevertheless, the cleaning work is relatively easy to perform. Consequently, plate and frame modules are now used only in some electrodialysis and pervaporation systems and in a limited number of reverse osmosis and ultrafiltration applications with highly fouling conditions [33].

2.2.2. Spiral-Wound

The spiral-wound module is characterized by a relative high packing density (>900 $m^2.m^{-3}$) and a simple design. Essentially, two or more 'membrane pockets' are wound around a permeate collecting tube with a particular mesh being used as spacers. The membrane 'pocket' covering two membrane sheets, with a highly porous support material (permeate-side spacer) in between, which are glued together along three edges. The fourth edge of the pocket is attached to the collecting tube. Several such pockets are spirally wound around the collecting tube with a feed-side spacer placed between the pockets forming a so-called 'element.' Usually, several elements are assembled in one pressure vessel. The feed-side flow is strictly axial in most designs (UOP, Abocor, Film Tec) or enters at the cylindrical surface of the element and exits axially. In any case, the permeate flows through the porous support inside the 'pocket' along the spiral to the collecting tube. Like the hollow fiber module, the spiral wound module cannot be cleaned mechanically. The classic spiral wound module is characterized by cross-flow and accurate modeling must, therefore, take into account the two-dimensional nature of velocity, pressure and concentration distribution for both feed and permeate channels. All models considering distributions only on one side (permeate or feed chamber) and assuming constant values on the other side are of limited use [34, 35].

2.2.3. Hollow Fiber

In general, the hollow fiber module consists of a pressure vessel containing a bundle of individual fibers [36], The open ends of the usually hairpin-like bent fibers are 'potted' into a head plate (an exception to this design is a gas permeation module manufactured by Dow Chemical with straight hollow fibers potted into a head plate at both ends). The feed solution may flow radially or in parallel with respect to the hollow fibers. Since the permeate is collected at the open ends of the fibers, the parallel flow may be either cocurrent or countercurrent, depending on whether the permeate in the fibers flows in the direction of the feed flow or against it. The distinction is important since the high friction losses in the fibers affect the local trans-membrane pressure difference [37].

It can be noticed that the two last modules combining a high packing density and reasonable manufacturing costs are by far the most successful designs. As a result, most of today's membrane-based gas separations are performed in hollow fiber modules, with perhaps fewer than 20% being performed in spiral-wound modules [38]. It was indicated that one trend of the development of the membrane module is a move to larger ones in order to reduce the whole cost of the separation [39]. Figure 5 highlights the size evolution of of hollow fiber modules during the last 30 years.

2.3. Membrane Classifications

2.3.1. Graphene as Membrane Material

Carbon is an amazing element that forms a wealth of materials from polymers to diamond to graphite. In fact carbon is the stuff of life, making up all of the organic compounds necessary for life, including DNA and other biomolecules and biopolymers. Beyond the organic compounds more than 107 synthetic molecules have been created in labs across the world that rely on the strong and stable carbon-carbon bond [40]. In addition to a wealth of organic compounds that carbon forms it also forms a variety of pure carbon allotropes, different structural forms of an element. This is contrast to the organic compounds which are mostly carbon but are also made up of other elements such as hydrogen, oxygen, and nitrogen.

The ability for carbon to form so many different compounds comes from the fact that its 4 valence electrons have very similar energies, unlike other elements with 4 valence electrons such as silicon. The similar energy levels of the 4 valence electrons allow their wave functions to mix easily and hybridize. This unique hybridization ability sets carbon apart from other elements and allows it to form quasi-0D, quasi-1D, quasi-2D, and 3D structures. Some of the various allotropes are fullerenes (0D), carbon nanotubes (1D), graphite and graphene (2D), and diamond and amorphous carbon (3D). These carbon allotropes allow carbon to exhibit a wide range of physical properties. Figure 6 shows some of the most common carbon allotropes [41].

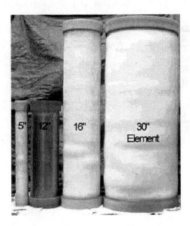

Figure 5. Photograph showing the development of hollow fiber modules at Cynara [30], from the first 5-in. modules of the 1980s to the 30-in. diameter behemoths currently being introduced. Photo courtesy of Cynara Company (now part of NATCO Group, Inc).

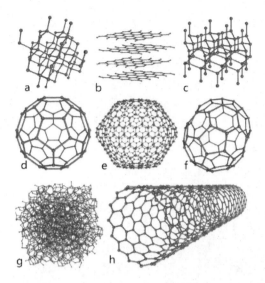

Figure 6. Carbon allotropes. (a) diamond, (b) graphite, (c) lonsdaleite, (d)-(f) fullerenes (C60, C540, C70) (g) amorphous carbon, and (h) carbon nanotube. Wikipedia.com: Allotropes of Carbon

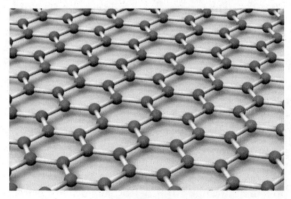

Figure 7. Graphene crystal structure. Carbon atoms are blue[1].

Graphene is a single layer of carbon atoms covalently bonded into a flat hexagonal lattice, a schematic of which is shown in Figure 7 When sheets of graphene are stacked together they form graphite, which is used in pencils

---

[1] Compliments of James Hedberg http://creativecommons.org/licenses/by-nc-sa/3.0/ http://www.jameshedberg.com/scienceGraphics.php?sort=all&id=graphene-atomic-structure-sheet

to write. The individual plans of graphene in graphite are bonded together by the weak van der Waals (vdW) force, as opposed to the strong covalent bonding in the plane. The weak bonding between sheets of graphene allows them to slide with respect to each other. This is what allows graphite to be used as a writing tool and also as a solid lubricant. The weak inter layer interaction is also what lead to the first isolation of graphene in 2004 by Konstantin Novoselov and Andre Geim [42].

Since its first isolation, graphene, as a research field has increased immensely, much like carbon nanotubes before it. The reason for the initial research surge into graphene was its electronic properties. Using graphene's electronic properties has led to a wide range of proposed applications. In addition to its superior electronic properties it is also nearly transparent, absorbing 2.3% of light per layer of graphene, which has led to additional applications in touch screen sensors and current collectors in solar cells [43].

Graphene is also an amazing material from a mechanical perspective. For being one atom thick, it is amazingly strong and has a Young's modulus (E) of ~1 TPa. Other mechanical properties that are important for graphene are its breaking stress σ_{int} and strain ε_{int} and its bending rigidity, B. It was found that graphene has a breaking stress of σ_{int} = 42 N/m and a breaking strain of ε_{int} = 25% [44]. Since graphene is two dimensional its bending rigidity is unlike that of ordinary materials where the bending rigidity is defined by the Young's modulus and the thickness of the material. In continuum mechanics the bending rigidity originates from simultaneous stretching and compression of the deformed material. Since graphene is only one layer of atoms the continuum model breaks down because there is not a top and bottom surface to bend and stretch. Instead the resistance to bending is an intrinsic property of the material that arises from interactions of the π and σ bonds. It should also be noted that the thickness of a monolayer of graphene cannot be unambiguously defined in the continuum sense [45]. As a result the bending rigidity of graphene is much less than the value derived from continuum mechanics [46, 47, 48]. Through molecular dynamics simulations the bending rigidity is estimated to be B ≈ 1-2 eV [49, 50] for single layer graphene and is 2 orders of magnitude lower than two and few layer graphene [46].

Not long after its initial isolation, Bunch et al. experimentally demonstrated that graphene, in its pristine form, is impermeable to all standard gas atoms including helium [51]. Soon after, theoretical papers began to emerge to explain the origin of graphene's impermeability [52]. The impermeability of graphene is attributed to its high crystal quality, low defect density, and the fact that the electron density of graphene's aromatic rings is large enough to repel atoms and molecules trying to pass through [52, 53]. Due to its impermeability, pores must be made in the graphene lattice to allow for selective transport of molecules.

The first theoretical study on graphene as a separation membrane was on the selective transport of ions across nanopores in a graphene sheet with different functional groups terminating the pore [54]. This study showed the viability of graphene as an ion separation membrane showing selectivity to anions or cations depending on pore functionality. Jiang et al. followed this up with the seminal theoretical study of porous graphene as a membrane for gas separations [55]. Jiang et al. studied both the permeability and selectivity of both nitrogen and hydrogen terminated pores in graphene and showed remarkable potential of graphene as a gas separation membrane for the separation of H_2 and CH_4. Figure 8a shows the mixed nitrogen and hydrogen terminated pore and Figure 8c shows the pore size due to the electron cloud. Their results showed an amazing H_2 permeance of 1 mol m^{-2} s^{-1} Pa^{-1} for a porous graphene membrane with a selectivity of 108 for H_2/CH_4. This result is a 107 improvement in permeance and a 105 to 107 improvement in selectivity over state of the art silica membranes. Additionally Jiang et al. calculated an even higher selectivity for their all H-terminated pore of 1023 (Figure 8b & d). A number of other studies have shown similarly remarkable gas separation properties of porous graphene with pores of vary sizes and functionalization [56, 57, 58, 59]. Perhaps even more remarkably, recent theoretical studies have shown that porous graphene has the ability to separate isotopes of helium and hydrogen gasses [60, 61, 62, 63].

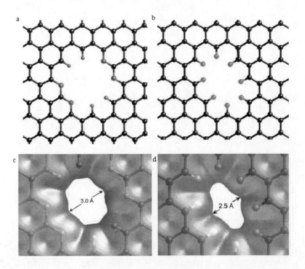

Figure 8. Graphene pores with mixed nitrogen (green) and hydrogen (blue) termination (a) and all hydrogen terminated pore (b). Equivalent pore size from electron density calculations for nitrogen and hydrogen terminated pore (c) and pure hydrogen terminated pore (d) [63].

2.3.2. Graphene-based Membranes

Graphene, Graphene Oxide (GO) and reduced Graphene Oxide (rGO), have been fabricated into membranes and shown promising performance in both gas and liquid separations due to their unique "size sieving" effect. Both nano-scaled interlayer spacing between two individual flakes and selective structural defects on the flakes have been claimed to be responsible for the observed separation performance.

Nair et al. prepared free-standing GO membranes by spray- or spin-coating of GO dispersion on porous substrates and subsequent transfer to a copper foil with an open hole [62]. They found that sub-micrometer thick GO membranes were completely impermeable to organic vapors and He gas, but allowed unimpeded permeation of water; the proposed water permeation pathway was interlayer spacing between GO flakes [64]. To elucidate the underlying mechanism, the authors reduced a GO membrane by annealing at 250°C in a hydrogen-argon mixed atmosphere. It exhibited 100 times less water flux, which was attributed to the narrowed interlayer distance from 1 to 0.4 nm (measured by X-ray diffraction (XRD)). Although He gas cannot permeate through dry GO membranes, it permeated through GO membranes with the existence of saturated water vapor. XRD results showed that when exposed to water vapor, water molecules can intercalate into the interlayer spacing due to the strong affinity between the oxygen groups and water and thus swell these capillaries; expanded interlayer spacing, therefore, allows He to permeate through. These capillaries also allow low-friction flow of water molecules, while blocking organic molecules. GO membranes, therefore, may have great potential of selective removal of water from organics. This work, for the first time, suggests GO membranes may separate different molecules through interlayer spacing between GO sheets. However, in this preliminary study, very thick GO membranes were prepared and studied for transport of molecules. The major advantage of using graphene-based material for preparing membranes, atomic thickness, therefore, was not realized. Koenig et al. investigated permeation of different gas molecules through individual porous graphene flakes. Graphene flakes, mechanically exfoliated from graphite, were suspended over micrometer-sized wells etched into silicon oxide wafer. They used ultraviolet-induced oxidative etching to controllably introduce defects into the pristine graphene flakes. A pressurized blister test and mechanical resonance were used to scale the transport of a range of gases (H_2, CO_2, Ar, N_2, CH_4 and SF_6) through the defects, and a molecular sieving treatment was realized. This proof-of-concept work demonstrates great potential of utilizing porous graphene as a promising membrane material for gas separation by molecular sieving. However, only graphene flakes, instead of membranes, were fabricated and tested in this study. Also, investigation of mixture gas separation wasn't performed.

In 2013, a group of scientist firstly prepared ultrathin, graphene-based membranes and demonstrated their gas separation performance. They fabricated ultrathin GO membrane with thickness approaching 1.8 nm by a facile vacuum filtration process on anodic aluminum oxide (AAO). Single gas permeation was first tested for He, H_2, CO_2, O_2, N_2, CO, and CH_4 molecules through an 18-nm GO membrane, and H_2 permeance was found to be nearly 300 times faster than CO_2. Afterwards, mixture gas separation was conducted and the selectivity was as high as 3400 and 900 for H_2/CO_2

and H_2/N_2 mixtures, respectively. Moreover, they observed that the H_2 and He permeances reduced exponentially as the membrane thickness increased from 1.8 to 180 nm, which could be due to the particular molecular transport pathway through the selective structural defects on the GO flakes. This conclusion was further strengthened by gas permeation through 18-nm rGO membranes (reduced d-spacing). Similar behavior as 18-nm GO membranes was observed, exhibiting that interlayer spacing is not the main transport pathway. This study suggests that ultrathin GO membranes may have wide applications. In a similar study, Kim et al. from Hanyang University in South Korea also investigated the gas permeation through ultrathin GO membranes [65]. They found GO membranes in the dry state were not permeable to gasses. However, water molecules intercalated in the GO interlayer spacing, which generated nanometer-sized pores and channels. These opened channels allowed permeation of gas molecules. The authors realized that the gases can permeate through thick GO membranes when sufficient pressure was exerted in order to overcome the energy barrier for pore entry and diffusion. Also, the gas permeability could be tuned by changing the GO flake size. They also demonstrated that a high CO_2/N_2 selectivity could be achieved when they varied the humidity levels in the feed streams of the GO membranes. Therefore, GO membranes may potentially be used to capture carbon dioxide from flue gas [66].

Celebi et al. reported a reliable technic for creating 2D graphene membranes utilizing CVD optimized to grow graphene with minimal defects and cracks to form graphene layers thinner than 1 nm [67]. Using a focused ion beam (FIB), they drilled nanopores in double layers of graphene to produce porous membranes with aperture diameters between less than 10 and 1000 nm. They found that the graphene membranes had water permeance five to seven times higher than conventional filtration membranes and water vapor flux was several hundred times higher than today's most advanced breathable textiles. This finding may lead to the development of highly breathable filters that are waterproof and effective to separate dangerous gases from air.

2.3.3. Polymeric Membranes

The polymers used for fabricating membranes can be classified as glassy polymers, such as polysulfone or polyimides, and rubbery polymers, like PDMS, depending on the chain length and relaxation time (chain rigidity). While diffusivity and solubility are the main factors that determine the permeability of gases, the permselectivity in glassy polymers is governed by diffusivity differences. Therefore, glassy polymers can separate gas molecules by exploiting size exclusion due to different kinetic diameters [68, 69]. However, permselectivity in rubbery polymers is dominated by solubility, so the highly condensable large molecules, such as butanes or pentanes are more permeable than small gases like N_2 in rubbery polymers. This is called 'reverse selectivity' due to the counter-intuitive separation result. Rubbery polymers are widely used to separate organic vapors from non-condensable gases such as air, N_2 [70, 71].

The demand for developing high performance separation membranes in industry is emerging. However, biggest large obstacle to utilize the polymeric membranes in industrial applications is their unsatisfactory separation performance. Polymeric membranes have a trade-off relationship between selectivity and permeability in gas phase separations as embodied in Robeson's plot [72].

2.3.4. Inorganic (Zeolite) Membranes

Zeolites are 3-D nanoporous crystalline materials that can be fabricated into highly selective separation membranes utilizing their molecular sieving nanopores [73, 74]. Membranes of different zeolites such as MFI, FAU and LTA are now being advanced for separation applications [75]. Notwithstanding the two decades of development on zeolite membranes with its promising characteristics of molecular selectivity, and a few commercialized applications such as alcohol dehydration, zeolite membranes are presently not widely utilized in industry for gas separations due to the difficulties in economical synthesis of a defect-free continuous membrane on a large scale. The challenging issues in zeolite membranes, such as scale-up of membrane manufacture and control over defect formation are currently being addressed to realize the industrial applications of zeolite membranes [76].

2.3.5. Polymer/Inorganic Composite ("Mixed Matrix") Membranes

"Mixed matrix" membranes can be created by incorporating suitably selected inorganic particles in a polymeric membrane. By consisting of both the polymer matrix and inorganic fillers, mixed matrix membranes can combine the advantages of polymers, such as good processibility, and lower cost, as well as those of the inorganic materials, such as high selectivity and thermal/ mechanical stability [77].

Many scientists have incorporated various types of inorganic materials such as zeolites, clays, and other molecular sieves into polymer matrices. Mahajan et al. incorporated zeolite 4A in Matrimid [78]. Carbon molecular sieve was also incorporated in Matrimid [79]. Merkel et al. fabricated fumed-silica nanoparticle/PMP (4-methyl-2-pentyne) composite membranes to improve the selectivity and throughput of organic vapour [80]. Peinemann incorporated ZSM-5 in a PDMS membrane to improve selectivity using its molecular sieving effect [81]. Also, a MFI and carbon black/PDMS composite has been applied as a pervaporation membrane for alcohol water separation [82, 83, 84]. Another effort on mixed matrix membranes was to improve surface compatibility between polymer and zeolite particles. Grignard and solvothermal surface treatments have been developed for surface compatibility, and helped to improve membrane permselectivity [85, 86].

The above works on mixed matrix membranes are quite promising towards the goal of improving the selectivity and permeability of polymer-based membranes. However, the conventional isotropic inorganic fillers often require high particle loadings (>50 wt%) in composite membranes to obtain substantial improvement [87]. Also, the size (usually bigger than 200 nm) of the particles makes it challenging to fabricate the thin (sub-micron) membranes. Porous layered materials have been suggested to overcome the above limitations of isotropic inorganic fillers in mixed matrix membranes [88-90]. Figure 9 shows a schematic diagram of a composite membrane incorporating high aspect ratio layered materials.

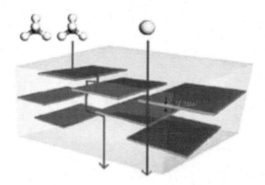

Figure 9. Schematic diagram of composite membranes including nanoporous platelets.

As shown in Figure 9, the high aspect ratio layered materials provides longer tortuous paths for larger molecules and effective to decrease the permeability of larger molecules [91]. The thinner shape of layered flakes may have lower diffusion limitation and larger adsorptive surface area than isotropic particles and possibly result in faster permeation for smaller molecules by comparing it to the thicker isotropic particles.

More importantly, utilization of the thin layered flakes is interesting because they can be incorporated in ultra-thin membranes such as the skin layer of hollow fiber membranes [92].

Beyond the utilization of layered materials in composite membranes, thin layered materials can potentially contribute to the formation of high quality inorganic membranes as well. As mentioned earlier, the current fabrication method for inorganic membrane using hydrothermal growth and calcination is having critical problems on defect formation and expensive pricings. So, direct deposition (coating) of layered materials in porous support can be the alternative way to prevent defect formation during high temperature process. This fabrication method by coating process may have the advantage on control of the membrane thickness by different coating conditions.

2.3.6. Layered Oxide Materials

Porous layered materials have created recent interest as molecular sieves and heterogeneous catalysts, because they can be exfoliated/delaminated into single-layered (or few-layered) nanostructures, and can provide many acid sites with a large surface area [93, 94, 95].

2.3.6.1. Clays

A clay is a representative non-porous layered material. The single sheet of clay has the thickness dimension of 1 nm and the crystal structure includes two tetrahedral sheets at the outer surface and 1 inner octahedral sheets. The difference valance between Al and Mg atoms in octahedral layer creates negative charges as active sites for catalytic applications [96]. Pinnavaia et al. used it as a heterogeneous catalyst by intercalating it with a metal complex, and it shows significant improvements on catalytic properties [97]. Also the structure of clay had been modified to micro/meso porous materials by a pillaring step [98] and the exfoliated clay had been incorporated in polymer matrix to improve the mechanical/chemical properties of engineering plastics [99, 100]. And lots of effort has also been made to improve barrier properties using the clay materials [101]. By combining it with fuel cell technology, there are active researches on DMFC membranes to decrease methanol crossover by putting exfoliated clay materials in Nafion® membranes [102, 103, 104].

2.3.6.2. MCM-22

MCM-22 is a porous layered material with a 10-membered ring (MR) pore structures with 2 lateral direction and 6MR pores along the vertical direction [105]. Corma et al. reported the swelling methods for MCM-22 using surfactant intercalation and exfoliated it to single layers to obtain a new layered material named ITQ-2 [106]. Different kinds of exfoliated materials such as ITQ-6 and ITQ-18 have been developed by diversifying the initial layered materials to Ferrierite and Nu-6, respectively [107, 109]. Also, MCM-36 has been introduced by pillaring MCM-22 after swelling process [106]. Exfoliated MCM-22 (ITQ-2) is completely attractive for high performance heterogeneous catalysis with its active sites in large surface area. Also, it is very useful as a molecular sieve for H_2 separations with its small pore openings of 6-membered ring perpendicular to the layers. The direct deposition of MCM-22 flakes has been used as the H_2 separation membranes [109].

2.3.6.3. AMH-3

AMH-3 is the first 3-D porous layered silicate/layered zeolite, having 8MR pores in all three principal crystallographic directions [110].

AMH-3 is an attractive candidate as selective flakes because the 8MR pore openings in AMH-3 crystals allow its application in various important gas separations. Table 4 shows the *Lennard-Jones* kinetic diameters of several gases. The nominally 3.4 Å sized pores of AMH-3 can be applied to various combinations of industrial gas separations such as H_2/CO_2 and CO_2/CH_4.

**Table 4. Lennard-Jones kinetic diameters (in Ångstrom)
of various [111]**

Gas	He	H_2	CO_2	O_2	N_2	CH_4
Kinetic diameter (Å)	2.6	2.89	3.3	3.46	3.64	3.8

A number of challenges have been addressed to exert this new material in genuine gas separation membranes. To overcome the problems on AMH-3 swelling process caused by the strongly bound Na^+ and Sr^{2+} between layers, a sequential intercalation method has been developed and the structural changes occurring in swollen AMH-3 have been elucidated [112, 113].

2.3.6.4. Layered MFI

Layered MFI is the most recently proposed nanoporous layered material. Choi et al. invented this material by synthesizing a new-conceptual templating agent that has long alkyl chain (C22) in one side [114]. A long chain alkyl of the template prevents the crystal growth to the b-direction. So, a 2 nm thick layered MFI could be synthesized in one-step hydrothermal reaction. Layered MFI has the same pore size and local structure as regular MFI. It has 10 MR pores with a pore size of 5.5 Å [115]. This pore size is larger than that of AMH-3, and can be exerted to the separation of larger molecules such as hydrocarbons or other organic molecules.

This 'already swollen' status of layered MFI is excellent advantage for utilization of this material in actual applications (e.g., catalysis or separation membrane). Because swelling processes usually involve tricky chemical reaction processes such as an ion exchange, intercalation by ammonium, etc.

By controlling the Na^+ content in the reaction gel, two different kinds of layered MFI, multilamellar and unilamellar, have been synthesized. The pillaring progress for layered MFI has expanded its potential functionalities in heterogeneous catalysis [116].

2.3.6.5. Porous Alluminophosphates (AlPOs)

Layered aluminophosphate (AlPO) materials are a new class of synthetic AlPOs that have potential applications in catalysis [117] and in fabrication of nanocomposite materials and membranes. By varying structure directing agent (SDA), the AlPO materials can be synthesized to the different forms such as, 1D (chain), 2D (layer), 3D open framework [118, 119, 120]. The dimensionality, pore structure, and composition of the AlPO are influenced by the size and shape of the organic (usually amine) SDAs [120]. Within those varieties, the Layered AlPOs are anionic and their Al:P ratio are lower than unity while the 3-D microporous of AlPOs is having Al:P stoichiometry of 1. The anionic AlPOs contain AlO_4 tetrahedra and $O = PO_3$ (or PO_4) units, with corner-shared oxygen atoms, to form AlPO materials with varying Al/P ratios [121, 122, 123].

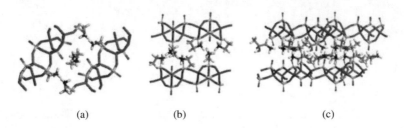

(a)   (b)   (c)

Figure 10. AlPO materials depending on different SDAs: (a) AlPOtrimethylamine (b) AlPO-triethylamine (c) AlPO-Isopropanolamine .

Figure 10 shows the molecular structure of layered AlPO materials depending on different SDA placed between each layers. The SDA molecules (such as trimethyl-, triethyl-, isopropanolamine) interact with AlPO framework *via* non-covalent bonding, and hence the 2-D layer structures can potentially be delaminated to single layers and enable new applications as catalysts or fabrication of nanocomposite membranes with its large active surface area. I have also contributed to this area by successful synthesis of a layered silicoaluminophosphate (SAPO) material that may be promising for hydrogen separations.

3. MODELING AND SIMULATION

3.1. Introduction

The main challenge for devising nanoporous membranes for gas separation is facing to adversity of manufacturing sub-nanometer pores with precisely-controlled sizes *via* experimental techniques which are necessary for molecular size exclusion. Various techniques have been put forward for manufacturing pore defects such as oxidation [124], ion bombardment [125, 126, 127, 128, 129] and electron-beam irradiation [130, 131], but assessments of gas transport rates through nanoporous membranes is not presented till now. Therefore, molecular modeling could be an essential procedure for illuminating the effect of pore size and chemistry on gas transport across nanoporous membranes.

Some molecular modeling and simulation works [132, 133, 134] have significantly centralized on quantum mechanics-based methods which distinguish the electronic structures of atoms in order to comprehend intermolecular interactions. Jiang et al. for instance implemented density functional theory (DFT) calculations to clarifies and dispute that nanoporous graphene has the potential to become the "ultimate" membrane for gas separation. The potential energy surface and dynamics of H_2 and CH_4 molecules passing cross sub-nanometer pores constructed in a graphene sheet was modeled utilizing both the Perdew, Burke, and Ernzerhof (PBE)

[135] functional form of the generalized-gradient approximation (GGA) and the Rutgers-Chalmers van der Waals density functional (vdW-DF) [136, 137] for exchange and correlation [138], the latter of which accounts for dispersion interactions. The PBE approach was consolidated with quantum mechanics-based First Principles Molecular Dynamics (FPMD) simulations, which indicated that by appropriately tuning nanopore sizes to obtain filtration by size exclusion, graphene can provide selectivity on the order of 10^8 to 10^{23} for H_2/CH_4 while exhibiting a H_2 permeance of 1 mol/m^2.s.Pa. This is up to 5 orders of magnitude larger than traditional polymeric membranes [139, 140], making it questionable, because it is notably larger than the equilibrium molecular flux of dilute H_2 on a membrane (0.2 mol/m^2.s.Pa); this discordance is even a greater amount for other molecular species. This clarifies the restrictions of DFT, and quantum-based methods principally - however these calculations are extremely useful for investigating the energetics of atomic interactions, they are now too computationally drastic to investigate gas transport in devices. In the study by Jiang et al. [138, 139], only 15 gas molecules were simulated, of which 4 crossings were considered and utilized to calculate the permeance. These numbers are very small and most feasibly inadequate to predict transport properties; hence, it is not surprising that their result is questionable.

Even though previous simulations [140, 141] have shown that surface adsorption realized by diffusion to the pore contributes to the flux through the pore, this was revealed to be small at the temperatures of interest here. At the temperatures of interest, gas molecules have high enough kinetic energy to directly overcome the energy barrier at the pore opening; hence surface adsorption and diffusion procedures are less substantial. Schrier [55] carried out highly computationally-intensive calculations at the MP2 level of theory using ccpVTZ basis sets to model the potential energy surface of He atoms passing through subnanometer pores created in a graphene sheet, and study the role of quantum and transmission effects as a function of temperature. At very low temperatures, classical transmission is exceptionally unlikely because there are very few atoms in the high energy tail of the Boltzmann distribution, while quantum tunneling procedures are orders of magnitude more likely to happen. However, at about 300 K, quantum effects account for only a 16% increase in the transmission of ^{4}He, and the increment further decreases at higher temperatures. Although, at 300 K, quantum effects measure for only a 16% in the transmission of ^{4}He, and the enhancement further decreases at higher temperatures [142]. This conclusion provides explanation for overlooking quantum effects and employing classical Molecular Dynamics (MD) simulations for studying gas transport across nanoporous graphene membranes (at room temperature) [143]. Mont Carlo (MC) simulation is also a widely used method. Unlike molecular dynamics simulations, MC simulations are free from the restrictions of solving Newton's equations of motion. This freedom allows for cleverness in the proposal of moves that generate trial configurations within the statistical mechanics ensemble of choice. Although these moves may be nontrivial, they can lead to huge speedups of up to 10^{10} or more in the sampling of equilibrium properties. Specific Monte Carlo moves can also be combined in a simulation allowing the modeler great flexibility in the approach to a specific problem. In addition, Monte Carlo methods are generally easily parallelizable, with some techniques being ideal for use with large CPU clusters. Using the MC simulation Yue and Yang [144] investigated the influence of temperature, pressure, and composition on the adsorption of pure and binary mixture of CO_2 and benzene in silicalite. Goj et al. [145] simulated the adsorption of CO_2 and N_2, both as single component and as binary mixture, in three zeolites with identical chemical composition but differing pore structures. The other group of scientists [146] analyzed the adsorption of binary mixture of CH_4/CF_4 in silicalite utilizing a gravimetric method and compared with the prediction of the ideal-adsorbed solution theory (IAST) and a multisite Langmuir model on the basis of the adsorption of single components.

On the other hand, understanding the diffusion of molecules in a porous material is important, as the diffusion behavior in confined space is differently from bulk phase. Diffusion in pores can be categorized in a number of various systems depending on the pore diameters. In a large pore over 50 nm or more in diameter, collisions between the molecules occur more frequently than collisions with the pore wall. This type of diffusion is usually referred as molecular diffusion or Fickian diffusion. With decreasing pore size, the number of collisions with the wall increases. As the pore size becomes smaller than the mean free path (the average distance travelled by a molecule between two collisions), the diffusion behaves in Knudsen regime, which is normally for pores with diameter between 2 and 50 nm. At even smaller pore, in which the diameter is comparable to the molecular size and molecules continuously experience the interaction with the pore, configurationally diffusion occurs.

Various theoretical frameworks have been devoted for modeling and simulation of different gases and the separation performance of nanoporous membranes. The main frameworks for are classified into four classes which detailed here.

3.1.1. Genuine Kinetic Separation

Molecular dynamics (MD) consists of two main branches; classical MD and quantum mechanics based MD (first principal MD). These two methods can provide genuine diffusion rate on a short time scale *via* calculating the internal forces. The units of these types of calculations are ns^{-1} Pa^{-1} (total flux) for a non-equilibrium molecular dynamics (NEMD) system that contains pressure, or molecules ns^{-1} for an equilibrium molecular dynamic (EMD) system without containing pressure difference through the membranes. This class of methods is appropriate for large systems with large number of gas molecules and the membrane frameworks. The other defection of this method while investigating carbon diffusion related to the existence of side-processes outside the determined timescale, which is confined to diffusive flux greater than about 10^{-12} m^2 s^{-1}. This difficulty is not limited to nanoporous carbon materials [147].

3.1.2. Statistical Kinetic Separation – Quantum Description

Statistical kinetic is based on statistical rate theory and specific 'bottleneck' points. This procedure stochastically simulates the kinetic separation fulfillment of nanoporous materials along the potential energy surface and does not simulate the gas diffusion with real time dynamics. Transition State Theory (TST) simplifies the approximate incorporation of quantum dynamical effects. One of the advantages of this method is much longer timescale simulation. There is an excellent between gas diffusion trends calculated by TST with those predicted using the calculated crossing times. DFT calculations are usually applied to achieve the Minimum Energy Pathway (MEP) for an individual gas molecule passing across the pores on the membrane. The common method for obtaining the MEP is to identify the most stable adsorption configurations of gases above the pore and inside the pore center. The latter can be obtained by transition state search algorithms, e.g., nudged elastic band (NEB) or Linear Synchronous Transit (LST), which are relatively extensively used in first principle electronic structure calculation software. Within this process many different configurations might be identified, and screening by frequency analysis and adsorption energy comparisons is significantly substantial. An actual transition state through a pore on a single-atomic layer membrane has one and only one imaginary frequency associated with motion across the membrane or pore, perpendicular to its plane. After the stationary points are identified, the reaction procedure can then be achieved by NEB or steepest-descent methods. Using partition functions, TST calculates the stationary points to generate rate constants for passage across through these supposedly rate-determining parts of the membrane. Rate constants will be embedded into stochastic kinetics simulation algorithm, resulting in to predict the diffusion rates and hence the selectivity with a great set of accuracy and possibility [148, 149].

3.1.3. Equilibrium Separation

In this method the equilibrium amount of each gas by the framework will be calculated. In this calculation on selective gas adsorption inside framework structures are generally implemented by Grand Canonical Monte Carlo (GCMC) molecular simulations. Its alternations consists of improved technologies such as Feynman–Hibbs effective potential (FH-GCMC) and the strict path integral method (PI-GCMC) [150, 151, 152].

3.1.4. Quantum Tunneling Corrections at Low Temperatures

In general, some studied systems contain an obstacle toward the diffusion of a gas molecule. In these cases quantum tunneling could effect on the behavior of the system at low temperatures where a classical statements predict no gas diffusion. The transmission probability $t(E)$ can be estimated as a function of the particle energy and then thermally weighted to achieve the thermal transmission probability of particles

$$p_w(T) = \int p(E,T)t(E)dE \qquad (6)$$

In such approach, a Boltzmann distribution for the 1D kinetic energies $p(E,T)$ of both species is accepted. As a whole, this effect allows more tunneling of lighter species than heavier species, showing opposite phenomenon with the quantum sieving mechanism [153,154].

Among these diverse methodological approaches, various structural models (cluster models and slab models) were utilized based on the dimension of the system and the level of chemistry chose. At the same time, different methods (molecular dynamics, *ab initio,* DFT with and without dispersion correction, etc.) were implemented due to the targeted properties.

3.2. Molecular Sieving

Molecular sieving is widely used in adsorptive separations. The separation mechanism in molecular sieving processes is kinetic in nature. In the case of air separation, for example, there is little difference in the equilibrium adsorption of the two principal components (oxygen and nitrogen) and the separation is due to the difference in the diffusion rates of these species through the pore network. Oxygen diffuses more rapidly than nitrogen; diffusion coefficient ratios between roughly 3 and 30 have been reported. This difference in the diffusion rates is easily understood. Lennard-Jones parameters for the atomic collision diameters are 0.294 nm for oxygen and 0.330 nm for nitrogen, giving a difference in size of about 10%. These species have similar diffusion coefficients in the bulk, but one would expect their diffusion coefficients to differ in pores that are close to their molecular dimensions. Because of its smaller size, oxygen would be expected to move more rapidly than nitrogen in a small pore [155]. This type of sieving mechanism is familiar from the study of zeolites, which have a crystalline structure. The sieving mechanism in a carbon-based molecular sieving (CMS), which is only locally crystalline and therefore contains pores of a range of sizes, is more complex. For a microporous carbon or other type of sieves to function as a molecular sieve, it is clear that most of the diffusing molecules must, at some point in their journey to the interior of the adsorbent, encounter constrictions of the right size to exert a sieving effect. For a system of pores in a series, the smallest pore is the dominant resistance to diffusion so that, provided this pore is of the appropriate size, molecular sieving results. The pore network of a real CMS, in contrast, is highly interconnected forming a series-parallel arrangement of resistances to diffusion. Thus, in a real CMS, each diffusing molecule has a very large number of possible routes from the surface of the adsorbent to its interior, each of which is composed of pores of a range of sizes; a "resistances in series" picture is inadequate for this type of structure. Perfect graphene is impermeable even to helium, however if we can make holes in graphene, we could make ideal helium-selective membranes [156]. The other widely investigated example of this effect is H_2 purification since the notable difference in its molecular diameters in comparison with other undesirable gases such as CO_2, CH_4 and, CO that often accompany hydrogen in industrial steam methane reforming [157]. Hydrogen as a high quality, clean energy carrier has attracted researcher attention. In the hydrogen production process, separation and purification technologies are considered to be the most critical issue. Thus the development of membrane separation technology could play an instrumental role in successful, economical production of molecular hydrogen [71]. One of the directed approaches to separate H_2 is to use a specifically designed molecular sieve with high selectivity and permeability. The main problem contributed to the traditionally utilized polymeric membranes is the trade-off between selectivity and permeability. The investigation on novel sieves with new structures to improve the permselectivity of H_2 from mixtures is essential for the development and utilization of hydrogen in a clean-energy economy. The very tiny molecular radius of the hydrogen molecule (the smallest except for helium) makes it suitable for investigating and exploiting the size sieving effect, especially in application to its mixture with CH_4 due to the large difference in their kinetic radii (for H_2 this value is 2.9 Å, and for CH4 3.8 Å). The first report of H_2/CH_4 separation by two dimensional carbon membranes was the nitrogen functionalized porous graphene *via* top-down approach was about the frameworks which are obtained by progressively varying the adsorption height and computing the corresponding energies [57]. The selectivity for H_2/CH_4 is predicted to be on the order of 10^8 with a high H_2 permeance. A FPMD study was performed to dynamically describe the hydrogen diffusion behavior on such a carbon membrane. Less than 600 K, the diffusion rate of 0.1 molecules ps^{-1} was observed, whiles no passage of CH_4 was observed during the total investigated time scale of 36 ns. A same structure constructed by passivating all hanging carbon bonds after pore creation was also investigated in the same paper.

As a whole, the size sieving effect by a carbon membrane with small pores could be utilized for the separation of small gas molecules (e.g., H_2 and He) from larger gas molecules. This size exclusion effect induced by the small pore size should be applied on gas pairs with considerable diameter difference, e.g., H_2/CH_4, or even H_2/N_2 and H_2/CO. The method would not be applicable for gas pairs with similar atomic diameter such as N_2 and CO. We should note that the size sieving effect is that for gas pairs with comparable sizes, the diffusion barrier does not necessarily observe the order of the kinetic diameters of gas molecules. For example, CO has a larger kinetic diameter (3.76 Å) than that of N_2 (3.64 Å), but the diffusion barrier across rhombic graphyne for the former (1.55 eV) is lower than that for the latter (1.73 eV). A similar relationship has also been reported for interaction of CO and N_2 with porous graphene synthesized via a bottom-up approach. These outputs highlight the importance of different chemical and physical interactions between the molecules and the pore in characterizing the diffusion barriers. By extrapolation, these results suggest that a size sieving effect could also be used in the separation of H_2 or helium from its mixtures with other large gas molecules in many industrial applications, including the separation of H_2/organic gases and organic gas mixtures such as n-C_4H_{10}/i-

C_4H_{10}, benzene/cyclohexane, and propylene/propane, as well as in sea water desalination [58, 158,159, 160, 161, 162, 163, 164].

4. QUANTUM SIEVING

4.1. Equilibrium Simulation of Quantum Sieving

Since the phenomenon of quantum sieving was first stated by Beenakker et al. [165] separation of H_2 isotopes via quantum sieving has attracted many of attentions. Wang et al. [166] were the first to affirm the existence of quantum sieving effects at 20 K in single-walled carbon nanotubes (SWCNTs) and interstices *via* the path integral grand canonical Monte Carlo (PI-GCMC) simulation. The consequences elucidate that SWCNTs with a pore size of less than 7 Å will exhibit high selectivity and T_2 will adsorb in suitable nanotubes and interstices where H_2 can be effectively excluded. Challa et al. [166, 167] also utilized PI-GCMC simulations to investigate the selectivity both in interstices and SWCNTs, considering the effects of temperature, pressure and pore size. Their results clarify that the selectivity of T_2/H_2 with an order of 10^5 in a (3, 6) tube and 10^4 in an (10, 10) interstice at 20 K is almost 20 times higher than D_2/H_2 under the same conditions. They revealed that the lowest selectivity occurred in a (6, 6) tube and the selectivity was significantly affected by the temperatures and pressures [168].

Since PI-GCMC simulations are time-consuming, Garberoglio used Boltzmann bias grand canonical Monte Carlo (BMC) simulations to appraise the selectivity in SWCNTs and carbon slit pores [169, 170]. He realized that significant quantum sieving effects exist both in SWCNTs and carbon slit pores and the selectivity exhibits a strong dependence on the temperature. The selectivity of D_2/H_2 is about 150 in carbon slit pores with pore widths of 5.7 Å at 20 K which drops to 3 at 77 K. The authors [171] also used the PI-GCMC simulations to study the effect of pressure and temperature on the selectivity in various organic framework materials. Their results show that the greatest quantum sieving effect is for the sorbent with the narrowest pores and the selectivity has a strong dependence on temperature while a less pronounced dependence on pressure. Moreover, the selectivity of T_2/H_2 in these organic frameworks is orders of magnitude lower than the extraordinary values predicted for narrower nanotube [168,169, 170, 171]. Recently, Liu et al. [172] introduced a new concept of 'quantum effective pore size' (QEPS), that is, an effective pore size measured by considering the swelling of the adsorbate–adsorbent potential size parameters caused by quantum effects and they apperceived good correlation between selectivity and QEPS. They found that two MOFs of $Cu(F-pymo)_2$ and CPL-1 exhibit especial selectivity which is higher than other MOFs as well as the porous materials such as carbon nanotubes, slit-shaped graphite and zeolites have been investigated so far [173]. Tanaka et al. [174] have expanded the selectivity of D_2/H_2 in (10, 10) interstices at 77 K *via* FH- GCMC simulation. D_2 molecules are preferentially adsorbed in (10, 10) interstices from the D_2/H_2 binary mixtures and the difference of adsorptive capacities between H_2 and D_2 decreases with increasing pressures. The researchers [175] studied the adsorption of D_2/H_2 and HD/H_2 binary mixtures in the carbon slit pores and cylindrical pores using the PI-GCMC and FH-GCMC simulations. The results showed that both approaches demonstrate good consistency and the quantum effects in the cylindrical pores are larger than the quantum effects in the slit pores. The optimal pore size for the separation of the D_2/H_2 and HD/H_2 mixtures is 0.623 and 0.625 nm respectively. The selectivity of D_2/H_2 at 77 K in a cylindrical pore with a pore width of 0.623 nm is 9.9 and the selectivity of HD/H_2 in a cylindrical pore with a pore width of 0.625 nm is 4.3.

4.1.1. Kinetic Simulations of Quantum Sieving

Increasing attention has been focused on kinetic quantum sieving for separating H_2 isotopes in recent years. In this section, recent progresses in separating H_2 isotopes *via* kinetic quantum sieving simulations will be presented. Bhatia and coworkers [172, 173, 176, 177] are one of the representatives who recognized the effectiveness for separating H_2 isotopes by the kinetic quantum sieving. Kumar et al. [176, 177, 178] observed reverse kinetic molecular sieving in zeolite-ρ at low temperatures for the first time in which they applied atomistic MD simulations, incorporating quantum effects *via* Feynman–Hibbs approach. The results revealed that zeolite- ρ with a window diameter of 0.596 nm does not show a reverse kinetic selectivity while it will do with a window diameter of 0.543 nm and the flux selectivity obtains the value of 22.66 at 30 K while staying rather low at higher temperature. In recent researches, Smith and co-workers [179] investigated the kinetic model for the transport of H_2 molecules in CMS *via* TST simulations to elucidate the effect of the pore-mouth and cavity on kinetic sieving. The results indicated that the differences in the length and layer of pore-mouth cannot affect the outcome of kinetic sieving while a small change in diameter (0.6 Å) will greatly affect the selectivity. Moreover, it is very attractive to note that the conclusions for the diffusion barriers in the cavity are in related to the model, means, nanotube-shaped or spherical-shaped models. The activation barrier lies in the cavity for nanotube-shaped models and in the pore-mouth for the spherical-shaped models. Therefore, scientists should consider the cavity shape and size in addition to the pore-mouth size in the next study for ensuring the rate-determining steps and achieving the quantum mediated kinetic molecular sieving of H_2 and D_2 on microporous materials.

4.1.2. Chemical Affinity Sieving

As a promising system for D_2/H_2 separation, nanoporous materials have attracted considerable research interest, as they can directly capture heavier isotope gas molecules via kinetic quantum sieving (KQS) [180, 181, 182, 183] or chemical affinity quantum sieving (CAQS) [183, 184, 185] . Heavier isotopes, having a shorter de Broglie wavelength at cryogenic temperatures, experience a lower diffusion barrier and can diffuse faster than lighter isotopes in a confined space, resulting in enrichment of the product with heavier isotope materials using the KQS effect [186]. Hirscher et al. enforced KQS effects by decorating the internal surface of metal-organic frameworks (MOFs) and covalent organic frameworks (COFs) with Cl^- and pyridyl groups, respectively, to optimize pore aperture, leading to significantly enhanced isotope selectivity ($S_{D2/H2}$) of 7.5 at 60 K and 9.7 at 22 K, respectively [182, 183]. However, due to the weak binding energies of hydrogen isotopes on the internal surface of porous materials, this approach effectively operates at only very low cryogenic temperatures. To overcome this limitation, CAQS exploits the preferential adsorption of heavier isotopes using strongly interacting sites introduced into the nanoporous materials [184, 185, 186]. As proposed theoretically as well as experimentally, the generation of open metal sites (OMSs) is a top priority for generating strong substrate interactions with H_2 and D_2 [187, 188, 189]. Fitz Gerald et al. first explored the CAQS effect of OMSs in D_2/H_2 separation showing a selectivity of 5 at 77 K, which demonstrated that efficient selectivity, can be obtained even at high temperatures using strong binding sites [185]. Nonetheless, a D_2/H_2 separation system comprising only OMSs is still far from meeting global demand, and thus there is a critical need for an effectively designed D_2-selective system. On the basis of previous findings, the ideal deuterium separation and storage system should have a high density of OMSs that can selectively adsorb deuterium even at high temperatures and a narrow pore size further increasing the separation effect through the selective diffusion of D_2 into the pores *via* KQS. In this sense, to realize very effective isotope separation, intelligently designed high capacity porous materials possessing both KQS and CAQS effects are required.

CONCLUSION AND OUTLOOK

Membrane technology has been considered as one of the most promising candidates for selective removal of specific gases from various mixtures. This chapter commenced with the brief history of gas separation and the evolutionary steps over the recent decades. This science was initiated in the early nineteenth century. Two essential criteria contributed to the significant development of this science, first, the extent diversity of membranes, particularly novel nanoporous membrane such as graphene and graphene-based membranes and second, wide applications of this science in industries and medicine. In the first section, different mechanisms of gas separation were discussed. The next section was dedicated to the introduction of nanoporous membranes. In this respect, formation of membranes and their structures were comprehensively considered. Final part focused modeling and simulation methods and recent investigations were briefly reviewed [190]. The current state of knowledge in this area has been advanced progressively and many researchers in all over the world are endeavoring to investigate and create novel computational methods for membrane gas separation. In the latest research, in 2018, Qiao and his coworkers, reported computational study to screen 4764 computation-ready experimental metal-organic frameworks (CoRE-MOF) for the membrane separation of a ternary gas mixture ($CO_2/N_2/CH_4$). Combining Monte Carlo and molecular dynamics simulations, the adsorption, diffusion and permeation of the gas mixture were predicted. The microscopic insights and structure-performance relationships from this computational study can facilitate the development of new MOF membranes for the upgrading of natural gas [191]. Gilassi et al. developed a mathematical model to simulate a gas separation process using a hollow fiber membrane module. They introduced new numerical technique based on flash calculation. This model is utilized to simulate a natural gas purification process using a single unit to determine the required membrane separation area and CH_4 loss [192]. Meng et al. also explored theoretically the structural and mechanical properties of metal-free fused-ring polyphthalocyanine (H_2PPc) and halogenated H2PPc (F H_2PPc and Cl-H_2PPc) membranes, and the energy profiles for gaseous H_2, CO, CH_4, CO_2 and N_2 molecules adsorbing on and passing through these monolayers. This method can evoke greatly the potential in energy and environmental fields [193]. Hoorfar et al. developed a key tool called "MemCal" to support the simulation of membrane-based gas separation processes using Aspen HYSYS. By integrating "MemCal" with the simulator, users can simulate complex multi-component/multi-stage processes, perform sensitivity studies, and utilize the default HYSYS features to cost and optimize processes. Industrially, "MemCal" can be used to simulate the separation of air, biogas, natural gas, olefins-paraffins and other gases. The CO_2/CH_4 binary mixture separation was demonstrated in this work as the first and the simplest application of "MemCal" [194]. Our research group is also interested in this field. Ganji and co-author evaluated the performance of nanoporous hexagonal boron nitride (h-BN) monolayers for hydrogen purification *via* employing DFT and MD simulations. They demonstrated that the 1B-3N pore is a far superior membrane to other counterparts and showed a significant potential for applications in hydrogen purification, clean energy combustion, and the design of novel membranes for gas separation [195]. In other investigation, a group of research designed some models of porous graphene with different sizes and shape as well as employ double layers porous graphene for efficient CO_2/H_2 separation. The selectivity and permeability of gas molecules through various nanopores were investigated by using the reactive molecular dynamics simulation which considers the bond forming/breaking mechanism for all atoms. The mechanism of gas penetration through the sub-nanometer pore was presented for the first time. The accuracy of MD simulation results evaluated *via* valuable DFT calculations. The results exhibited that reactive MD simulation can propose an economical means of separating gases mixture [196]. Moradi et al. investigated in water purification performance of various SWCNT membranes with different diameters and opening chemistries by reactive MD simulations and DFT computations. The conclusion showed that nanotube diameter and opening chemistry of the membrane significantly affect the membrane purification performance. The CNT-based membrane presented in this research might be arranged in certain water purification as well as gas separation membranes that would support fast and efficient water purification (such as organic pollutants/water mixtures including phenol/benzene/cyclohexane and dioxin) and can resolve environmental treatments [197]. The challenges and further investigations can provide opportunities for in-depth molecular simulation studies toward the practical applications of membrane gas separation.

ACKNOWLEDGMENTS

The authors gratefully acknowledge support of this work by the Tehran Medical sciences, Islamic Azad University, Tehran, Iran.

REFERENCES

[1]   Adhikari, S.; Fernando, S. *Industrial & Engineering Chemistry Research* 2006, 45, 875.

[2]   Kikkinides, E. S.; Yang, R. T.; Cho, S. H. *Engineering Chemistry Research* 1993, 32, 2714.

[3]   Miller, G. Q.; Stoecker, J. *Selection of a hydrogen separation process.* National Petroleum Refiners Association, Washington, DC, 1989.

[4]   Hinchliffe, B.; Porter, K. E. *Chemical Engineering Research and Design* 2000, 78, 255.

[5]   Sircar, S.; Golden T. C. *Separation Science & Technology* 2000, 35, 667.

[6]   Cheryan M., *Ultrafiltration Handbook*, Taylor & Francis 1998.

[7]   Rautenbach, R.; Gröschl, A. *Desalination* 1990, 73, 84.

[8]   Byrne, W.; *Reverse Osmosis: A Practical Guide for Industrial Users*, Tall Oaks Pub., 2002.

[9]   Cath, T. Y.; Childress, A. E.; Elimelech, M. *Journal of Membrane Science* 2006, 281, 70.

[10] Feng, X.; Huang, R. Y. M.; L *Industrial & Engineering Chemistry Research* 1997, 36, 1048.

[11] Mansourizadeh, A.; Ismail, F. *Journal of Hazardous Materials* 2009, 171, 38.

[12] Kesting, R. E.; Fritzsche, A. K.; *Polymeric gas separation membranes*, John Wiley & Sons, New York, 1993.

[13] Koros, W. J.; Fleming, G. K. *J. Membr. Sci.* 1993, 83.

[14] Mulder, M. *Basic principles of membrane technology*, Dordrecht: Kluwer Academic Publishers, 1996.

[15] Henis, J. M. S.; Tripodi, M. K. *J. Membr. Sci.* 1981, 8, 233.

[16] Paul, D. R..; Yampol'skii, Y. P. *Polymeric Gas Separation Membranes* CRC Press, Boca Raton, 1994.

[17] Chung, T. S. *Polym. and Polym. Comp.* 1996, 4, 269.

[18] Kluiters, S. C. A. Status review on membrane systems for hydrogen separation., *5th Research Framework Programme of the European Union*, 2004.

[19] Koros, W. J. *Gas separation in membrane separation systems: Recent developments and future directions*, USA, 1991,189.

[20] Uhlhorn, R. J. R.; Keizer, K. A. J. *J. Membr. Sci.* 1989, 46, 225.

[21] Koresh, J.; Soffer, A. *Sci. Technol.* 1987, 22, 973.

[22] Way, J. D.; Roberts, D. L. *Sci. Technol.* 1991, 27, 29.

[23] Koros, W. J.; Fleming, G. K.; Jordan, S. M.; Kim, T. H.; Hoehn, H. H. *Prog. Polym. Sci.* 1988, 13, 339.

[24] Mahajan, R. *Formation, characterization and modeling of mixed matrix membrane materials*, Ph. D. Thesis, The University of Texas at Austin, 2000.

[25] Robeson, L. M. *J. Membr. Sc.* 1991, 62, 165.

[26] Berens, A. R.; Hopfenberg, H. B. *J. Membr. Sci.* 1982, 10, 283.

[27] Strathmann, H. *J. Membr. Sci.* 1981, 9, 121.

[28] Lloyd, D. R. Membrane materials science, an overview. In *ACS Symposium Series,* Washington D. C., USA, 1985.

[29] V. R. E. Kesting. *Synthetic polymeric membranes. A structural perspective. 2nd Edition.* John Wiley and Sons, N Y, 1985.

[30] Baker, R. W.; Lokhandwala, K. *Ind. Eng. Chem. Re.* 2008, 47, 2109,

[31] Spillman, R. W. *Chem. Eng. Prog,* 1989, 85, 41.

[32] Wang, L.; Corriou, J. P.; Castel, C.; Favre, E. Proc´ed´e de séparation membranaire en régime discontinu [Membrane separation process in batch mode]. *France Patent*: FR 11 60587. 2011.

[33] Wang, P.; Chung, T. S. *J. Memb. Sci.* 2015, 474, 39.

[34] Jain, M.; Attarde, D.; Gupta, S. K. *J. Memb. Sci.* 2016, 507, 43.

[35] Jain, M.; Attarde, D.; Gupta, S. K. *J. Memb. Sci.* 2015, 490, 328.

[36] Rautenbach, R.; Albrecht, R.; *Membrane Processes*. John Wiley & Sons, Chichester, 1989.

[37] Baker, R. W. *Ind. Eng. Chem. Res.,* 2002, 41, 1393.

[38] Baker, R. W. *Membrane Technology and Applications*. Wiley, Chichester, 2004.

[39] Li, Y.; Zhou, Z.; Shen, P.; Z. Chen. *Chemical Communications* 2010, 46, 3672F. Diederich.; Y. Rubin, Synthetic Approaches toward Molecular and Polymeric Carbon Allotropes, *Angewandte Chemie International Edition in English.* 1992,31, 1101.

[40] Dembinski, R.; Bartik, T.; Bartik, B.; Jaeger, M.; Gladysz, J. A. *J. Am. Chem. Soc.* 2000, 122, 810.

[41] Geim, K.; Kim. P. *Scientific American* 2008, 298, 90.

[42] Bonaccorso, F.; Sun, Z.; Hasan, T.; Ferrari, A. C. *Nature Photonics* 2010, 4, 611.

[43] Lee, C.; Wei, X.; Kysar, J. W.; Hone, J. *Science* 2008, 321, 385.

[44] Huang, Y.; Wu, J.; Hwang, K. C.; *Physical Review B.* 2006, 74, 245413.

[45] Zhang, D. B.; Akatyeva, E.; Dumitrică, T. *Physical Review Letters* 2011, 106, 255503.

[46] Lu, Q.; Arroyo, M.; Huang, R. *Journal of Physics D: Applied Physics* 2009, 42, 102002.

[47] Tapasztó, L.; Dumitrică, T.; Kim, S. J.; Nemes-Incze, P.; Hwang, C.; Biró, L. P. *Nature Physics* 2012, 8, 739.

[48] Wang, C.; Mylvaganam, K.; Zhang, L. *Physical Review* B, 2009, 80, 155445.

[49] Wei, Y.; Wang, B.; Wu, J.; Yang, R..; Dunn, M. L. *Nano Letters* 2013, 13, 26.

[50] Bunch, J. S.; Verbridge, S. S.; Alden, J. S.; van Der Zande, A. M.; Parpia, J. M.; Craighead, H. G.; McEuen, P. L.; Van Der Zande, A. M. *Nano Letters* 2008, 8, 2458.

[51] Leenaerts, O.; Partoens, B.; Peeters, F. M. *Applied Physics Letters* 2008, 93, 193107.

[52] Jiang, D.; V. Cooper, R.; Dai, S. *Nano letters 2009, 9, 4019.*

[53] Sint, K.; Wang, B.; Král, P. *Journal of the American Chemical Society* 2008, 130, 16448.

[54] Du, H., Li, J.; Zhang, J.; Su, G.; Li, X.; Y. Zhao. *Journal of Physical Chemistry C* 2011, 115, 23261.

[55] Schrier, J. *The Journal of Physical Chemistry Letters* 2010, 1, 2284

[56] Jiao, Y.; Du, A.; Hankel, M.; Smith, S. C. *Phys. Chem. Chem. Phys.* 2013, 15, 4832.

[57] Blankenburg, S.; Bieri, M.; Fasel, R.; Müllen, K.; Pignedoli, C. A.; Passerone, D. *Small* 2010, 6, 2266.

[58] Schrier, J.; McClain. J.*Chemical Physics Letters* 2012, 521, 118.

[59] Hauser, W.; Schwerdtfeger, P. *The Journal of Physical Chemistry Letters* 2012, 3, 209.

[60] Hankel, M.; Jiao, Y.; Du, A.; Gray, S. K.; Smith, S. C. *The Journal of Physical Chemistry C.* 2012, 116, 6672.

[61] Hauser, W.; Schrier, J.; and Schwerdtfeger, P. *The Journal of Physical Chemistry C.* 2012, 116, 10819.

[62] Nair, R. R.; Wu, H. A.; Jayaram, P. N.; Grigorieva, I. V.; Geim, A. K. *Science* 2012, 335, 442.

[63] Jiang, D. E.; Cooper, V. R.; Dai, S. *Nano Lett.* 2009, 9, 4019.

[64] Koenig, S. P.; Wang, L.; Pellegrino, J.; Bunch, J. S. *Nature nanotechnology* 2007, 728-732.

[65] Li, H.; Song, Z. N.; Zhang, X. J.; Huang, Y.; Li, S. G.; Mao, Y. T.; Ploehn, H. J.; Bao, Y.; Yu, M. *Science* 2013, 342, 95.

[66] Kim, H. W.; Yoon, H. W.; Yoon, S. M.; Yoo, B. M.; Ahn, B. K.; Cho, Y. H.; Celebi, H. J. K.; Buchheim, J.; Wyss, R. M.; Droudian, A.; Gasser, P.; Shorubalko, I.; Kye, J. I.; Lee, C.; Park, H. G. *Science,* 344, 28.

[67] Stern, S. A.; Mi, Y.; Yamamoto, H.; Stclair, A. K. *J Polym Sci Pol Phys.* 1989, 27, 1887.

[68] Park, J. Y.; Paul, D. R. *J Membrane Sci* 1997, 125, 23.

[69] Schultz, J.; Peinemann, K. V. *J Membrane Sci* 1996, 110, 37.

[70] Pinnau, I.; He, Z. J. *J Membrane Sci* 2004, 244, 227.

[71] Robeson, L. M. *J Membrane Sci* 2008, 320, 390.

[72] Davis, M. E. *Nature* 2002, 417, 813.

[73] Snyder, M. A.; Tsapatsis, M. *Angew Chem Int Edit* 2007, 46, 7560.

[74] Caro, J.; Noack, M. *Micropor Mesopor Mat* 2008, 115, 215.

[75] Choi, J.; Jeong, H. K.; Snyder, M. A.; Stoeger, J. A.; Masel, R. I.; Tsapatsis, M. *Science* 2009, 325, 590.

[76] Koros, W. J. *Aiche J* 2004, 50, 2326.

[77] Mahajan, R.; Koros, W. J. *Polym Eng Sci* 2002, 42, 1420.

[78] Vu, D. Q.; Koros, W. J.; Miller, S. J. *J Membrane Sci* 2003, 211, 311.

[79] Merkel, T. C.; Freeman, B. D.; Spontak, R. J.; He, Z.; Pinnau, I.; Meakin, P.; Hill, A. J. *Science* 2002, 296, 519.

[80] Jia, M. D.; Peinemann, K. V.; Behling, R. D. *J Membrane Sci* 1991, 57, 289.

[81] Vankelecom, I. F. J.; DeKindercn, J.; Dewitte, B. M.; Uytterhoeven, J. B. *J Phys Chem B* 1997, 101, 5182.

[82] Moermans, B.; De Beuckelaer, W.; Vankelecom, I. F. J.; Ravishankar, R.; Martens, J. A.; Jacobs, P. A. *Chem Commun* 2000, 2467.

[83] Vane, L. M.; Namboodiri, V. V.; Bowen, T. C. *J Membrane Sci* 2008, 308, 230.

[84] Shu, S.; Husain, S.; Koros, W. J. *J Phys Chem C* 2007, 111, 652.

[85] Bae, T. H.; Liu, J. Q.; Lee, J. S.; Koros, W. J.; Jones, C. W.; Nair, S. *J Am Chem Soc* 2009, 131, 14662.

[86] Adams, R. T.; Lee, S. S.; Bae, T. H.; Ward, J. K.; Johnson, J. R.; Jones, C. W.; Nair, S.; Koros, W. J. *J Membrane Sci* 2011, 367, 197.

[87] Jeong, H. K.; Krych, W.; Ramanan, H.; Nair, S.; Marand, E.; Tsapatsis, M. *Chem Mater* 2004, 16, 3838.

[88] Zornoza, B.; Gorgojo, P.; Casado, C.; Tellez, C.; Coronas, J. *Desalin Water Treat* 2011, 27, 42.

[89] Galve, A.; Sieffert, D.; Vispe, E.; Tellez, C.; Coronas, J.; Staudt, C.:

[90] Copolyimide mixed matrix membranes with oriented microporou. *J Membrane Sci* 2011, 370, 131.

[91] Ray, S. S.; Okamoto, M. *Prog Polym Sci* 2003, 28, 1539.

[92] Johnson, J. R.; Koros, W. J. *J Taiwan Inst Chem E* 2009, 40, 268.

[93] Dasgupta, S.; Torok, B. *Org Prep Proced Int* 2008, 40, 1.

[94] Liu, P.: Polymer modified clay minerals. *Appl Clay Sci* 2007, 38, 64.

[95] Centi, G.; Perathoner, S.*Micropor Mesopor Mat* 2008, 107, 3.

[96] Paul, D. R.; Robeson, L. M. *Polymer* 2008, 49, 3187.

[97] Pinnavaia, T. J. *Science* 1983, 220, 365.

[98] Kloprogge, J. T. *J Porous Mat* 1998, 5, 5.

[99] Usuki, A.; Kojima, Y.; Kawasumi, M.; Okada, A.; Fukushima, Y.; Kurauchi, T. Kamigaito, O. *J Mater Res* 1993, 8,1179.

[100] Messersmith, P. B.; Giannelis, E. P.*Chem Mater* 1994, 6, 1719.

[101] Choudalakis, G.; Gotsis, A. *European Polymer Journal* 2009, 45, 967.

[102] Rhee, C. H.; Kim, H. K.; Chang, H.; Lee, J. S. *Chem Mater* 2005, 17, 1691.

[103] Thomassin, J. M.; Pagnoulle, C.; Bizzari, D.; Caldarella, G.; Germain, A.; Jerome, R.. *Solid State Ionics* 2006, 177, 1137.

[104] Lin, Y. F.; Yen, C. Y.; Hung, C. H.; Hsiao, Y. H.; Ma, C. C. M. *J Power Sources* 2007, 168, 162.

[105] Leonowicz, M. E.; Lawton, J. A.; Lawton, S. L.; Rubin, M. K. *Science* 1994, 264, 1910.

[106] Corma, A.; Fornes, V.; Pergher, S. B.; Maesen, T. L. M.; Buglass, J. G. *Nature* 1998, 396, 353.

[107] Corma, A.; Diaz, U.; Domine, M. E.; Fornes, V. *J Am Chem Soc* 2000, 122, 2804.

[108] Corma, A.; Fornes, V.; Diaz, U. *Chem Commun* 2001, 2642.

[109] Choi, J.; Tsapatsis, M. *J Am Chem* Soc 2010, 132, 448.

[110] Jeong, H. K.; Nair, S.; Vogt, T.; Dickinson, L. C.; Tsapatsis, M. *Nat Mater* 2003, 2, 53.

[111] Robeson, L. M. *J Membrane Sci* 1991, 62, 165.

[112] Choi, S.; Coronas, J.; Jordan, E.; Oh, W.; Nair, S.; Onorato, F.; Shantz, D. F.; Tsapatsis, M. *Angew Chem Int Edit* 2008, 47, 552.

[113] Choi, S.; Coronas, J.; Sheffel, J. A.; Jordan, E.; Oh, W.; Nair, S.; Shantz, D. F.; Tsapatsis, M. *Micropor Mesopor Mat* 2008, 115, 75.

[114] Choi, M.; Na, K.; Kim, J.; Sakamoto, Y.; Terasaki, O.; Ryoo, R. *Nature* 2009, 461.

[115] Diaz, I.; Kokkoli, E.; Terasaki, O.; Tsapatsis, M. *Chem Mater* 2004, 16, 5226.

[116] Na, K.; Choi, M.; Park, W.; Sakamoto, Y.; Terasaki, O.; Ryoo, R. *J Am Chem Soc* 2010, 132, 4169.

[117] Thomas, J. M.; Raja, R.; Sankar, G.; Bell, R. G. *Accounts of Chemical Research* 2001, 34, 191.

[118] Li, J. Y.; Yu, J. H.; Yan, W. F.; Xu, Y. H.; Xu, W. G.; Qiu, S. L.; Xu, R. R. *Chem Mater* 1999, 11, 2600.

[119] Williams, I. D.; Yu, J. H.; Gao, Q. M.; Chen, J. S.; Xu, R. R. *Chem Commun* 1997, 1273.

[120] Yu, J.; Sugiyama, K.; Zheng, S.; Qiu, S.; Chen, J.; Xu, R.; Sakamoto, Y.; Terasaki, O.; Hiraga, K.; Light, M.; Hursthouse, M. B.; Thomas, J. M. *Chem Mater* 1998, 10, 1208.

[121] Yan, W. F.; Yu, J. H.; Shi, Z.; Xu, R. R. *Chem Commun* 2000, 1431.

[122] Yu, J.; Terasaki, O.; Williams, I. D.; Quiv, S.; Xu, R. *Science* 1998, 5, 297.

[123] Williams, I. D.; Gao, Q. M.; Chen, J. S.; Ngai, L. Y.; Lin, Z. Y.; Xu, R. R. *Chem Commun* 1996, 1781.

[124] Shirley, J. H. *Phys. Rev.* 1965, 138.

[125] Ferey, G.; Latroche, M.; Serre, C.; Millange, F.; Loiseau, T.; Percheron-Guegan, A. *Chem. Commun* 2003, 2976.

[126] Liu, J.; Culp, J. T.; Natesakhawat, S.; Bockrath, B. C.; Zande, B.; Sankar, S. G.; Garberoglio, G.; Johnson, J. K. *J. Phys. Chem.* C 2007, 111, 9305.

[127] Ferey, G.; Mellot-Draznieks, C.; Serre, C.; Millange, F.; Dutour, J.; Surble, S.; Margiolaki, I. *Science* 2005, 309, 2040.

[128] Cote, A. P.; Benin, A. I.; Ockwig, N. W.; O'Keee, M.; Matzger, A. J.; Yaghi, O. M. *Science* 2005, 310, 1166.

[129] El-Kaderi, H. M.; Hunt, J. R.; Mendoza-Cortes, J. L.; Cote, A. P.; Taylor, R. E.; O'Keee, M.; Yaghi, O. M. *Science* 2007, 316, 268.

[130] Furukawa, H.; Yaghi, O. M. *J. Am. Chem. Soc.* 2009, 131, 8875.

[131] Park, K. S.; Ni, Z.; Cote, A. P.; Choi, J. Y.; Huang, R. D.; Uribe-Romo, F. J.; Chae, H. K.; O'Keee, M.; Yaghi, O. M. *Proc. Natl. Acad. Sci. USA.* 2006, 103, 10186.

[132] Huang, X.-C.; Lin, Y.-Y.; Zhang, J.-P.; Chen, X.-M. *Angew. Chem. Int. Ed.* 2006, 45, 1557.

[133] Arean, C. O.; Manoilova, O. V.; Bonelli, B.; Delgado, M. R.; Palomino, G. T.; Garrone, E. *Chem. Phys. Lett.* 2003, 370, 631.

[134] Du, X. M.; Wu, E. D. *Chin J Chem Phys* 2006, 19, 457.

[135] Schimmel, H. G.; Kearley, G. J.; Mulder, F. M. *ChemPhysChem* 2004, 5, 1053.

[136] Moller, C.; Plesset, M. S. *Phys. Rev.* 1934, 46, 618.

[137] Hohenberg, P.; Kohn, W. *Phys. Rev.* 1964, 136, B864.

[138] Hedin, L.; I. Lundqvist, B.; Lundqvist, S. *Solid State Commun.* 1971, 9, 537.

[139] Du, X. M.; Wu, E. D. *Chin J Chem Phys* 2006, 19, 457.

[140] Singh, S.; Eijt, S. W. H.; Huot, J.; Kockelmann, W. A.; Wagemaker, M.; Mulder, F. M. *Acta Mater.* 2007, 55, 5549.

[141] Perdew, J. P.; Burke, K.; Ernzerhof, M. *Phys. Rev. Lett.* 1996, 77, 3865.

[142] Perdew, J. P.; Burke, K.; Wang, Y. *Phys. Rev. B* 1996, 54, 16533.

[143] Schimmel, H. G.; Kearley, G. J.; Mulder, F. M. *Chem Phys Chem* 2004, 5, 1053.

[144] Yue, X. P.; Yang, X. N. *Langmuir 2006*, 22, 3138.

[145] Goj, A.; Sholl, D. S.; Akten, E. D.; Kohen, D. *J. Phys. Chem. B* 2002, 106, 8367.

[146] Buss, E.; Heuchel, M. J. *Chem. Soc., Faraday Trans.* 1997, 93, 1621.

[147] Voter, A. F.; Montalenti, F.; Germann, T.C. *Annu Rev Mater Res.* 2002, 32, 321.

[148] Henkelman, G.; Jonsson, H. *J. Chem. Phys.* 2000, 113, 9978.

[149] Halgren, T. A.; Lipscomb, W. N. *Chem. Phys. Lett.,* 1977, 49, 225.

[150] Wang, Q.; Johnson, J. K.; Broughton, J. Q. *J. Chem. Phys.* 1997, 107, 5108.

[151] Tanaka, H.; Noguchi, D.; Yuzawa, A.; Kodaira, T.; Kanoh, H.; Kaneko, K. *J. Low Temp. Phys.* 2009, 157.

[152] Garberoglio, G.; Johnson, J. K. *ACS Nano* 2010, 4, 1703–1715.

[153] Ratner, M. A.; Schatz, G. C. *An introduction to quantum mechanics in chemistry,* Prentice Hall, Upper Saddle River, NJ, 2000.

[154] J. I. Steinfeld, J. S. Francisco and W. L. Hase, Chemical kinetics and dynamics, Prentice Hall, *Upper Saddle River*, NJ, 1999.

[155] Chen, Y. D.; Yang, R. T.; Uawithya, P. *AIChE J.* 1994, 40, 577.

[156] Steele, W. A. *The Interaction of Gases with Solid Surfaces*; Pergamon Press: Oxford, 1974.

[157] Lu, G. Q.; da Costa, J. C. D.; Duke, M.; Giessler, S.; Socolow, R.; William, R. H.; Kreutz, T. *J. Colloid Interface Sci.* 2007, 314, 589.

[158] Poot, M.; van der Zant, H. S. J. *Applied Physics Letters* 2008, 92, 63111. Zhang, H.; He, X.; Zhao, M.; Zhang, M.; Zhao, L.; Feng, X.; Luo, Y.*J. Phys. Chem. C.* 2012, 116, 16634.

[159] Perdew, J. P.; Wang, Y. *Matter Mater. Phys.* 1992, 45, 13244.

[160] De Vos, R. M.; Henk, V. *Science* 1998, 279, 1710.

[161] McKeen, L. W. *Permeability Properties of Plastics and Elastomers*, Elsevier, 3rd edn, 2012.

[162] Lee, H. R.; Kanezashi, M.; Shimomura, Y.; Yoshioka, T.; Tsuru, T. *AIChE J.* 2011, 57, 2755.

[163] Lee, H. R.; Shibata, T.; Kanezashi, M.; Mizumo, T.; Ohshita, J.; Tsuru, T. *J. Membr. Sci.* 2011, 383, 152.

[164] Cohen-Tanugi, D.; Grossman, J. C. *Nano Lett.* 2012, 12, 3602.

[165] Beenakker, J. J. M.; Borman, V. D.; Krylov, S. Y. *Chem. Phys. Lett.* 1995, 232, 379.

[166] Wang, Q.; Challa, S. R.; Sholl, D. S.; Johnson, J. K. *Phys. Rev. Lett.* 1999, 82, 956.

[167] Challa, S. R.; Sholl, D. S.; Johnson, J. K. *Condens. Matter* 2001, 63, 245411.

[168] Challa, S. R.; Sholl, D. S.; Johnson, J. K. *J. Chem. Phys.* 2002, 116, 814.

[169] Garberoglio, G. *Eur. Phys. J. D* 2009, 51, 185.

[170] Garberoglio, G. *Chem. Phys. Lett.* 2009, 467, 270.

[171] Liu, D.; Wang, W.; Mi, J.; Zhong, C.; Yang, Q.; Wu, D. *Ind. Eng. Chem. Res.* 2011, 51, 434.

[172] Wang, Y.; Bhatia, S. K.; *J. Phys. Chem. C.* 2009, 113, 14953.

[173] Wang, Y.; Bhatia, S. K. *Mol. Simul.* 2009, 35, 162.

[174] Tanaka, H.; Noguchi, D.; Yuzawa, A.; Kodaira, T.; Kanoh, H.; Kaneko, K. *J. Low Temp. Phys.*, 2009, 157, 352.

[175] Tanaka, H.; Miyahara, M. T.; *J. Chem. Eng. Jpn.* 2011, 44, 355.

[176] Kumar, V. A.; Bhatia, S. K. *Phys. Rev. Lett.* 2005, 95, 245901.

[177] Kumar, V. A.; Jobic, H.; Bhatia, S. K. *J. Phys. Chem. B,* 2006, 110, 16666.

[178] Kumar, V. A.; Jobic, H.; Bhatia, S. K. *Adsorption,* 2007, 13, 501.

[179] Kumar, V. A.; Bhatia, S. K. *J. Phys. Chem. C,* 2008, 112, 11421.

[180] Hankel, M.; Zhang, H.; Nguyen, T. X.; Bhatia, S. K.; Gray, S. K.; Smith, S. C. *Phys. Chem. Chem. Phys.*, 2011, 13, 7834.

[181] Teufel, J.; Oh, H.; Hirscher, M.; Wahiduzzaman, M.; Zhechkov, L.; Kuc, A.; Heine, T.; Denysenko, D.; Volkmer, D. *Adv. Mater.* 2013, 25, 635.

[182] Oh, H.; Kalidindi, S. B.; Um, Y.; Bureekaew, S.; Schmid, R.; Fischer, R. A.; Hirscher, M. *Angew. Chem., Int. Ed.* 2013, 52, 13219.

[183] Cai, J.; Xing, Y.; Zhao, X. *RSC Adv.* 2012, 2, 8579.

[184] Kowalczyk, P.; Terzyk, A. P.; Gauden, P. A.; Furmaniak, S.; Pantatosaki, E.; Papadopoulos, G. K. *J. Phys. Chem. C* 2015, 119, 15373.

[185] FitzGerald, S. A.; Pierce, C. J.; Rowsell, J. L.; Bloch, E. D.; Mason, J. A. *J. Am. Chem. Soc.* 2013, 135, 9458.

[186] Oh, H.; Savchenko, I.; Mavrandonakis, A.; Heine, T.; Hirscher, M. *ACS Nano* 2014, 8, 761.

[187] Savchenko, I.; Mavrandoaski, A.; Heine, T.; Oh, H.; Teufel, J.; Hirscher, M. *Microporous Mesoporous Mater.* 2015, 216, 133.

[188] Beenakker, J. J. M.; Borman, V. D.; Krylov, S. Y. *Chem. Phys. Lett.* 1995, 232, 379.

[189] Suh, M. P.; Park, H. J.; Prasad, T. K.; Lim, D. W. *Chem. Rev.* 2012, 112, 782

[190] Cheon, Y. E.; Suh, M. P. *Chem. Commun.* 2009, 2296.

[191] Qiao, Z.; Xu, Q.; Jiang, J. *J Memb Sci.* 2018, 551, 47.

[192] Gilassi, S.; Taghavi, S. M.; Rodrigue, D.; Kaliaguine, S. *AIChE J.* 2018, 64, 1766.

[193] Meng, Z.; Zhang, Y.; Shi, Q.; Liu, Y.; Du, A.; Lu, R. *Phys Chem Chem Phys.* 2018.

[194] Hoorfar, M.; Alcheikhhamdon, Y.; Chen, B. *Comput Chem Eng.* 2018, 117, 11.

[195] Ganji, M. D.; Dodangeh, R. *Phys Chem Chem Phys.* 2017, 19, 12032.

[196] Esfandiarpoor, S.; Fazli, M.; Ganji, M. D. *Sci Rep.* 2017, 7, 1. Moradi, F.; Ganji, M. D.; Sarrafi, Y. *Phys Chem Chem Phys.* 2017, 19, 8388.

In: Gas Separation
Editor: Suraya Mathews

ISBN: 978-1-53614-606-6
© 2019 Nova Science Publishers, Inc.

Chapter 2

NEW DEVELOPMENTS IN ALTERNATIVE FILLERS FOR MIXED MATRIX MEMBRANE SYNTHESIS

Oh Pei Ching[*], Pannir Selvam Murugiah,
Bong Yin Chung, Muhammad Asif Jamil
and Grace Tan Ying En
CO_2 Research Centre (CO_2RES),
Institute of Contaminant Management,
Chemical Engineering Department, Universiti Teknologi PETRONAS,
Bandar Seri Iskandar, Perak, Malaysia

ABSTRACT

Mixed matrix membrane (MMM) is a composite material comprising organic phase and inorganic fillers. The primary role of fillers is to systematically manipulate the molecular packing of the organic phase, thus enhance the gas separation properties of MMMs. For the past few decades, the emphasis was mainly placed on the incorporation of inorganic fillers

[*] Corresponding Author E-mail: peiching.oh@utp.edu.my.

such as zeolite, carbon molecular sieve, and silica. These fillers are commonly used due to their narrow pore size which can yield high selectivity. Nevertheless, it is often difficult to produce MMM with defect-free morphologies using these fillers due their poor distribution properties and interfacial adhesion with the polymer matrix, which deteriorate their gas separation performance. Consequently, research has expanded to alternative fillers such as layered silicate, carbon nanotube (CNT), graphene oxide (GO) polyhedral oligomeric silsesquioxane (POSS) and titanium dioxide (TiO$_2$). These alternative fillers are predominantly used in the preparation of nanocomposites with increased thermal, chemical and mechanical stability. These enhancements have improved the engineering capability of nanocomposites to be used in a wide range of application which includes elastomers, thermoset plastics, thermoplastics, materials for drug delivery, dental and medical applications. However, the use of alternative fillers for gas separation MMM synthesis is still limited. Carbonaceous nanofiller with organized carbon structure has been found to exhibit outstanding reinforcing properties on the polymer matrix. Despite its unique morphological and structural properties, limited research has been conducted using carbonaceous nanofiller for MMM synthesis. Thus far, the application of carbonaceous nanofillers such as carbon nanotubes (CNT) and graphene oxide (GO) has shown promising result in terms of gas separation performance. However, the separation efficiency of CNT-based MMMs can only be achieved if CNTs are aligned vertically. Likewise, the promising result by GO is challenged by the bulk quantity production of the graphene sheet. Recently, polyhedral oligomeric silsesquioxane (POSS) is receiving immense attention due to its monodispersed size, low density, and ease of modification. The truly unique feature of POSS lies in its inner inorganic and outer organic framework, which can be made up of various organic or inorganic substituents. The hybrid properties it possesses allows POSS to be dispersed homogeneously in the polymer matrix without agglomeration or aggregation. Hence, accompanied by the intrinsic molecular property of POSS, the performance of the POSS-based MMMs can be enhanced. Nonetheless, the performance of these MMMs is dictated by the incorporation technique employed during membrane fabrication. TiO$_2$ is an emerging non-porous metal oxide filler for membrane separation application. Its inherent advantage comes from its ability to be dispersed individually. The addition of optimum loading of metal oxides may lead to significant enhancement in the permeability of the membrane while maintaining or enhancing membrane selectivity. On the other hand, the overall permeability and selectivity may also decrease due to the non-porous nature of metal oxides which are unable to selectively sieve out molecules of different sizes. Much recent work has gone into attempting to tailor novel alternative fillers which could be incorporated into MMM for gas separation application.

Keywords: alternative fillers, carbonaceous nanofillers, polyhedral oligomeric silsesquioxane (POSS), clay minerals, titanium dioxide (TiO$_2$), gas separation, mixed matrix membrane

1. INTRODUCTION

The separation of gases by mixed matrix membrane (MMM) has received much attention in recent years due to its cost-effectiveness as well as the ability to exceed Robeson's permeability and selectivity upper bound limit (Liu et al. 2008). MMM consists of a continuous polymer phase and dispersed inorganic fillers. Theoretically, the dispersion of inorganic fillers in the polymer phase can create a preferential pathway for desired permeates and act as a barrier for undesired retentate (Jadav and Singh 2009). Besides that, inorganic fillers can also enhance MMM's mechanical, chemical and thermal stability while the polymer phase provides superior processability (Goh et al. 2011a). Despite its advantages, the successful development of MMM is highly dependent on the compatibility between the polymer phase and inorganic fillers.

Over the past few decades, much of the research conducted has focused on inorganic fillers such as zeolites, carbon molecular sieves (CMS) and silica. The apparent criterion for the selection of zeolites and carbon molecular sieves as the dispersed phase in MMM development is due to their narrow pore sizes that are able to yield high selectivity (Tewari 2016, Nasir et al. 2013). On the other hand, non-porous silica is added in MMMs for its ability to alter the polymer chain packing which increases its fractional free volume. This alteration without the creation of non-selective voids will result in the enhancement of membrane permeability and selectivity. Nevertheless, their poor interfacial adhesion with the polymer matrix proved to be detrimental to their gas separation performance (Bastani, Esmaeili, and Asadollahi 2013, Caro and Noack 2008, Husain and Koros 2007). Their popularity further declines as they require labor intensive synthesis procedures as well as their inability to simultaneously increase the permeability and selectivity of MMM.

Consequently, research has been expanded towards the incorporation of alternative fillers such as carbonaceous nanofillers, clay mineral, metal oxide and hybrid organic-inorganic material. Currently, carbon nanotube (CNT) and graphene oxide (GO) are two of the most prominent carbonaceous fillers used in membrane development research. CNT has since been deemed to exhibit the most attractive gas transport properties compared to any other materials due to its open-ended hollow structure and inherent smoothness. The addition of these nanotubes in MMM is able to enhance its mechanical strength while retaining its separation efficiency (Zhao et al. 2017, Ge et al. 2011). Likewise, graphene has become a new and emerging material for MMM development because of its high aspect ratio and inherently similar characteristics to CNT without the high synthesis cost (Ji et al. 2016). At the same time, pristine graphene has been found to be impermeable to molecules even as small as a helium atom. Instead, it is permeable to protons (Bunch et al. 2008, Tsetseris and Pantelides 2014). Although pristine graphene is of little use in membrane development, graphene oxide (GO), which is the oxidized form of graphene has been used in the development of MMM with superior gas separation performance.

Meanwhile, a new type of organic-inorganic material known as polyhedral oligomeric silsesquioxane (POSS) is also receiving immense attention as alternative filler. POSS possesses several attractive attributes such as monodispersed size, low density, and ease of modification. These characteristics allow POSS to be dispersed homogeneously in a polymer matrix without agglomeration or aggregation. Thus far, POSS-containing polymer nanocomposites have experienced enhancement in mechanical and thermal properties. This has led to their wider range of applications in elastomers, thermoset, thermoplastics, materials for drug delivery, dental and medical applications, materials possessing low dielectric, fire retardant, laser, space resistant, weather resistant, scratch and wear resistant properties, fuel cells, tire, and others. However, the application of POSS for the development of gas separation MMM remains relatively new. Past studies have demonstrated that the organic outer layer of POSS is able to improve its compatibility with a polymer matrix, thus, overcome the main challenge experienced by many conventional inorganic fillers.

Currently, clay minerals such as layered silicate, montmorillonite (MMT), halloysite nanotube (HNT), sepiolite and laponite have also been used as the dispersed phase in MMM development. Layered silicate which acts as a 'selective flake' is predicted to enhance MMM separation performance. This can be attributed to its high aspect ratio and a thin layer (Kim et al. 2013). High aspect ratio creates a highly tortuous path for larger molecules, decreasing their permeability while the thin layer allows an increased permeability of smaller molecules (Goh et al. 2011a). Layered silicates can be obtained naturally or synthetically. The factors influencing clay minerals-based MMM's permeability includes their alignment in the polymer matrix, aspect ratio, loading and delamination degree (Zulhairun and Ismail 2014a). Previous work has reported the use of synthetic layered silicates such as AMH-3, JDF-L1, and AIPO. Incorporation of AMH-3 in cellulose acetate polymer matrix has shown enhancement in CO_2 permeability without compromising CO_2/CH_4 selectivity (Kim et al. 2013). Nevertheless, the application of JDF-L1 and AMH-3 remains limited because their successful incorporation into a polymer matrix is dependent on their degree of swelling and exfoliation.

Titanium dioxide (TiO_2) is an emerging non-porous metal oxide filler for membrane separation application. Its inherent advantage comes from its ability to be dispersed individually. The addition of metal oxides at optimum loading may lead to two possible outcomes. Firstly, the permeability of the membrane can be significantly enhanced while maintaining or enhancing membrane selectivity (Moghadam et al. 2011, Liang et al. 2012). On the other hand, overall permeability and selectivity may diminish due to the non-porous nature of metal oxides, which is unable to selectively sieve molecules of different sizes (Moradihamedani et al. 2015).

2. CARBONACEOUS NANOFILLERS

For gas separation MMM, the addition of inorganic fillers in the polymer matrix is expected to alter the transport properties of gases, modify the properties of adjacent polymer phase and alter the packing, dynamics or conformation of polymer chains near its surface (Houshmand, Wan Daud, and Shafeeyan 2011). According to Ismail et al. the integration of inorganic fillers enhance the gas separation properties of the polymer matrix via two possible methods (Ismail et al. 2009a). Firstly, the interaction between polymer-filler particles disrupts the polymer chain packing, increases the polymer matrix free-volume and eventually increases the gas diffusivity parameter. Thereafter, the functional groups attached to the filler surface may interact with certain gases and enhance the gas solubility parameter. In addition, the incorporation of filler particles is also believed to improve the mechanical properties and thermal stability of the polymeric membrane. However, these enhancements are only possible if the filler particles are dispersed homogeneously in the polymer matrix without significant polymer-filler interfacial defects (Julian and Wenten 2012). To date, various inorganic particles have been incorporated into the polymer matrix to synthesize MMM. These inorganic fillers could be divided into (1) conventional fillers (i.e., zeolite, carbon molecular sieve, and silica) and (2) alternative fillers (i.e., carbon nanotubes, layered silicate, metal oxide, polyhedral oligomeric silsesquioxane and metal organic framework). Excessive yet significant works have been carried out with conventional fillers due to their unique and distinct properties which contributed to the robust performance of the resultant membrane. Nevertheless, Goh et al. found that the development of MMMs using conventional fillers have reached its bottleneck whereby the performance of resultant membrane could not be further improved to fulfill the increasing expectation for practical applications (Goh et al. 2011b). Various modifications have been performed on these filler particles and the performance of resultant membranes have been well documented. Compared to conventional filler, very limited work has been reported on the utilization of alternative fillers in MMM development. Alternative fillers especially carbonaceous nanofillers with unique morphological and structural properties are relatively new for gas separation membrane development in which its full potential is yet to be widely explored. In recent years, carbonaceous nanofillers emerged to be the most promising and sophisticated class of materials. They are well known for their organized carbon structure which exhibits outstanding reinforcing properties based on their assembly in the polymer matrix. They also have the ability to improve the engineering capability of the polymer matrix and has been studied in nearly every field of engineering and science. Carbon nanotube, carbon nanofiber, and graphene oxide are the common carbonaceous nanofillers used in nanocomposites applications.

2.1. Carbon Nanotube (CNT)

Since the discovery of carbon nanotube (CNT) in 1991 by Ijima, it has been widely studied in various applications due to its unprecedented mechanical, thermal and electrical properties (Mittal et al. 2015). CNT is one of the primary carbonaceous nanofillers used in polymer composites reinforcement. It is made up of hexagonal graphene sheets rolled up into a cylindrical form and capped with half a fullerene structure (Mittal et al. 2015). It could be categorized into single-walled carbon nanotube (SWCNT) and multi-walled carbon nanotube (MWCNT) depending on the layer of graphene sheets in the cylindrical shape, as depicted in Figure 2.1. Compared to SWCNT, MWCNT is cheaper and available in large quantities due to its advanced production technique (Qiu et al. 2010).

Based on the simulation work performed by Sholl et al. on the transport rate of gases through the CNT-based membrane, the gas permeability is found to be in orders of magnitude faster than any other inorganic fillers, surpassing zeolite-based membranes (Chen and Sholl 2006, Skoulidas, Sholl, and Johnson 2006). This was predicted to be due to the unique morphology, larger diameter, high aspect ratio and frictionless surface of CNT. Thus, gas particles are able to pass through the smooth walls of CNT with less collision, thereby increasing the diffusion rate of fast gas (Houshmand, Wan Daud, and Shafeeyan 2011). Besides its inherent properties, the interaction of CNT particles with the polymer matrix could disrupt the polymer chain packing, introduce more polymer matrix free volume and improve gas diffusion properties (Goh et al. 2011b). CNT is also capable of enhancing the mechanical properties of the polymer matrix even at a lower filler loading (Ismail et al. 2009b). Recent experimental works which incorporate CNT particles in a polymer matrix for gas separation MMM development are summarized in Table 2.1. From Table 2.1, it was noticed that the addition of CNT increases the gas permeability across the membrane but reduces the gas selectivity of the polymer matrix. This might result from the increase in polymer matrix free volume as well as polymer-filler interfacial defects which allow the gases to permeate faster without significant discrimination.

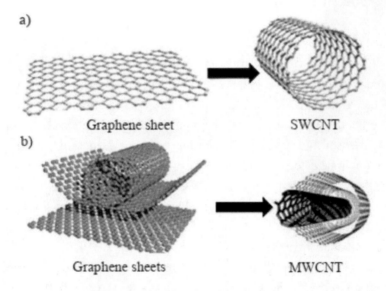

Figure 2.1. Schematic diagram of a) SWCNT and b) MWCNT.

In addition, CNT also possesses poor dispersibility properties in the polymer matrix (Li, Ma, et al. 2015a). CNT's large aspect ratio, smaller size, and strong π-π stacking effect is prone to agglomerate and forms tight bundles or hollow ropes in the polymer matrix (Mittal et al. 2015). Therefore, a series of pre-treatment either functionalization or purification are generally performed on CNF surface prior to MMM fabrication. This additional step is executed to debundle the highly entangled nanotubes into individual CNT for homogeneous dispersion in the polymer matrix as well as to improve the interfacial interaction with the polymer chain (Goh et al. 2011b). Filler functionalization could be divided into covalent and non-covalent functionalization. For covalent functionalization, CNT is usually treated with strong acids or oxidizing agents such as HNO_3, H_2SO_4, H_2O_2, and $KMnO_4$. The surface functionalization process introduces various functional groups such as hydroxyl, carboxyl, alcohol, and ketone at the open ends and defects side of CNT. These polar groups reduce the forces between CNT particles, disentangle or disintegrate the CNT bundles into individual particles, thus enhance the dispersion of the filler in the polymer matrix (Khabashesku 2011). Moreover, the newly introduced organic functionalities also generate a stronger interaction between the polymer chain and filler particles (Ge, Zhu, and Rudolph 2011).

Nonetheless, chemical functionalization could result in significant damage to the native CNT structure and deteriorates its mechanical and thermal properties (Mittal et al. 2015). Ismail *et al.* suggested that chemical functionalization of CNT must be carried out in a controlled condition to govern the degree of functionalization as well as to sustain the native properties of the CNT particles (Ismail et al. 2009b). However, it is often difficult and laborious to control the degree of chemical functionalization on the filler surface. Thus, non-covalent functionalization has been introduced to connect the molecules without forming any chemical bonds, hence do not affect to the native structure and properties of CNT. Polymer wrapping and surfactant dispersion are the most common non-covalent functionalization methods that have been applied for CNT. In the polymer wrapping method, the polymer will be wrapped helically around the CNT wall to form supermolecular complexes of CNTs. Meanwhile, in the surfactant dispersion method, the CNT is dispersed in surfactant with different charges (Goh et al. 2011b). Both of these processes are expected to reduce the CNT surface tension which inhibits filler agglomeration, thereby enhancing the dispersion of the filler in the polymer matrix (Ismail et al. 2009b).

Apart from poor dispersion properties, the fabrication of CNT-based MMM also encounters several other challenges (Li, Ma, et al. 2015b). The conventional membrane fabrication methods are ineffective in aligning the CNT in the polymer matrix as the nanotubes are exceptionally flexible and exhibit high aspect ratio (Sung et al. 2004). Besides that, CNT also needs to be purified via a multi-stage purification process due to the presence of metallic and amorphous carbon impurities (Kingston et al. 2004). If not adequately conducted, the CNT structure might be damaged.

Table 2.1. Gas separation performance of dense CNT-based MMM

Polymer	Feed pressure	Weight %	Performance of polymeric membrane	Performance of CNT-based MMM	Ref
$BPPO_{dp}$	10 psig	5 wt% SWCNT	$P_{CO_2}=$ 82 Barrer $\alpha_{CO_2/N_2}=$ 30	$P_{CO_2}=$ 129 Barrer $\alpha_{CO_2/N_2}=$ 29	(Cong et al. 2007)
		5 wt% MWCNT	$P_{CO_2}=$ 82 Barrer $\alpha_{CO_2/N_2}=$ 30	$P_{CO_2}=$ 148 Barrer $\alpha_{CO_2/N_2}=$ 32	
Matrimid	2 bar	10 wt% MWCNT	$P_{CO_2}=$ 8.84 Barrer $\alpha_{CO_2/CH_4}=$ 34 $\alpha_{CO_2/N_2}=$ 32.74	$P_{CO_2}=$ 10.29 Barrer $\alpha_{CO_2/CH_4}=$ 27.81 $\alpha_{CO_2/N_2}=$ 26.38	(Li, Ma, et al. 2015b)
PDMS	N/A	10 wt% SWCNT	$P_{CO_2}=$ 166.02 Barrer $\alpha_{CO_2/CH_4}=$ 5.90	$P_{CO_2}=$ 191. 30 Barrer $\alpha_{CO_2/CH_4}=$ 5.30	(Kim, Pechar, and Marand 2006)
PBNPI	2 kg/cm^2	15 wt% MWCNT	$P_{CO_2}=$ 2.61 Barrer $\alpha_{CO_2/CH_4}=$ 3.73	$P_{CO_2}=$ 6.00 Barrer $\alpha_{CO_2/CH_4}=$ 3.37	(Weng, Tseng, and Wey 2009)
PSf	4 bar	10 wt% SWCNT	$P_{CO_2}=$ 3.90 Barrer $\alpha_{CO_2/CH_4}=$ 23 $\alpha_{CO_2/N_2}=$ 23 $\alpha_{CO_2/O_2}=$ 4.7	$P_{CO_2}=$ 5.19 Barrer $\alpha_{CO_2/CH_4}=$ 19 $\alpha_{CO_2/N_2}=$ 23 $\alpha_{CO_2/O_2}=$ 4.3	(Kim et al. 2007)
PES	2 bar	5 wt% MWCNT	$P_{CO_2}=$ 2.60 Barrer $\alpha_{CO_2/N_2}=$ 21.60	$P_{CO_2}=$ 4.50 Barrer $\alpha_{CO_2/N_2}=$ 22.5	(Ge, Zhu, and Rudolph 2011)

2.2. Carbon Nanofiber (CNF)

Carbon nanofiber (CNF) is also a carbonaceous nanofiller which received considerable attention in recent years. CNF has been used to replace CNT as it exhibits similar properties with CNT at lower manufacturing cost (Al-Saleh and Sundararaj 2009). High production cost, poor dispersibility in the polymer as well as the tedious purification of CNT encourage the utilization of CNF in polymer nanocomposites development. Lately, CNF has been utilized in energy storage and conversion devices such as batteries, electrode materials, sensors, and supercapacitors, due to its excellent electrical, mechanical and thermal properties (Kingston et al. 2004). CNF has a unique structure, known as stacked-up carbon nanotubes in which the graphene layers are curved at an angle, α to create stacks of nanocones, as shown in Figure 2.2 (Tran, Zhang, and Webster 2009).

As illustrated in Figure 2.2, the graphene sheets are perfectly wrapped into a cylindrical structure to form CNT whereas they are stacked at approximately 25 degrees from its fiber axis to form CNF. Since the graphene plane edges are exposed in CNF, it could easily interact with the polymer matrix without any complex functionalization process. In addition to that, CNF is also well-known for its high compatibility with most polymer processing techniques, unlike CNT. Apart from its physical structure, other noticeable differences between CNT and CNF are summarized in Table 2.2.

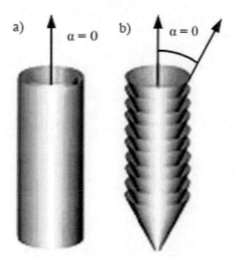

Figure 2.2. Schematic representation of a) CNT and b) CNF.

Similarly, the overall performance of CNF-polymer composites is dependent on the quality of CNF dispersion and its interaction with the polymer matrix (Feng, Xie, and Zhong 2014). Since the graphene plane edges are exposed in CNF, it can easily interact with the polymer chain. However, considerable ambiguity remains on the polymer-CNF bonding type, bonding strength and dislocation densities which requires further in-depth analysis. Generally, CNF is dispersed in the polymer matrix via (1) melt mixing process using extrusion or roll mill (Sánchez et al. 2011), Haake torque rheometer (Lozano, Bonilla-Rios, and Barrera 2001) and mini-max molder (Tibbetts, Finegan, and Kwag 2002) or (2) sonication process (Pervin et al. 2005). Melt mixing is a very common approach which could yield a uniform dispersion of CNF particles in the polymer matrix. Nevertheless, this technique requires high shear mixing condition which could reduce the aspect ratio of CNF and eventually deteriorate the final properties of the composites (Tibbetts et al. 2007, Lozano, Bonilla-Rios, and Barrera 2001, Al-Saleh and Sundararaj 2009). On the other hand, the sonication process is relatively simpler. It involves sonication of CNF in a solvent for a predetermined duration, followed by addition of polymer via mechanical stirring (Pervin et al. 2005). In this technique, external cooling might be required to control the temperature rise during the sonication process.

Table 2.2. Comparison between CNF and CNT

Filler	CNT	CNF
Size	1-30 nm	50-200 nm
Price	Expensive (Up to 750 dollars per gram)	Reasonable (0.22-1.10 dollars per gram)
Pre-Treatments	Required (Purification and Functionalization)	Optional (Simple purification)
Van der Waals forces	Very strong	Weaker than CNT
Dispersion in polymer matrix	Poor	Good
Physical Properties	Excellent	Good

Moreover, the surface treatment of CNF has also been attempted by several researchers to improve dispersion and interaction characteristics of CNF in the polymer matrix. In most studies, CNF surface is either oxidized with nitric/sulfuric acid or treated with amine groups, which is expected to act as the bridging-agent connecting CNF with the polymer chain, and hinders particle agglomeration within the composites (Li et al. 2005).

2.3. Graphene Oxide (GO)

Since Nobel Prize was awarded for graphene development in 2010, graphene has become the topic of interest in numerous research field (Hu et al. 2014). Graphene was also nominated as the strongest material with a Young modulus of TPa (Bolotin et al. 2008). It exhibits extremely impressive properties within a single material such as high thermal conductivity up to 5000 W/m.K (Balandin et al. 2008), high electron mobility 20,000 cm^2/V.s (Bolotin et al. 2008), good mechanical properties, high transparency, large specific surface area up to 2630 m^2/g, high aspect ratios (>1000), very low density and good compatibility with polymer (Johnson, Dobson, and Coleman 2015, Hu et al. 2014, Van Noorden 2006). Basically, graphene is made up of carbon compound in a 2D flat sheet structure and is synthesized via mechanical exfoliation of graphite or chemical vapor deposition method (Zhang et al. 2012). Nevertheless, the synthesis of defect-free graphene is difficult and time-consuming. Thus, after extensive studies have been conducted, it is found that graphene-based derivatives which preserve the properties of native graphene are easier to be synthesized. Out of the many derivatives, graphene oxide (GO), the oxidized form of graphene seems to be the most popular choice as it is able to retain the extraordinary properties of native graphene while exhibiting improved dispersibility and processability compared to the precursor (Konios et al. 2014).

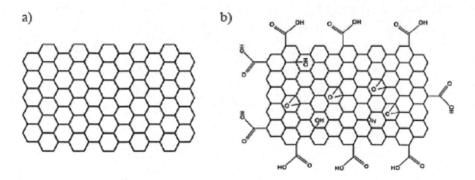

Figure 2.3. Schematic structure of a) graphene and b) graphene oxide.

Table 2.3. GO-based MMM for CO_2 separation application

Polymer	Feed pressure (bar)	Weight %	Performance of polymeric membrane	Performance of GO-based MMM	Ref
PES	5	10 vol%	$P_{CO_2}=$ 7.13 GPU $\alpha_{CO_2/CH_4}=$ 29.70	$P_{CO_2}=$ 17.59 GPU $\alpha_{CO_2/CH_4}=$ 30.32	(Ebrahimi et al. 2016)
Pebax	1	10 wt%	$P_{CO_2}=$ 480 Barrer $\alpha_{CO_2/CH_4}=$ 18	$P_{CO_2}=$ 220 Barrer $\alpha_{CO_2/CH_4}=$ 23	(Li, Cheng, et al. 2015)
Matrimid	2	10 wt%	$P_{CO_2}=$ 8.84 Barrer $\alpha_{CO_2/CH_4}=$ 34.00	$P_{CO_2}=$ 6.46 Barrer $\alpha_{CO_2/CH_4}=$ 70.30	(Li, Ma, et al. 2015b)
PEO	10	1.0 wt%	$P_{CO_2}=$ 280 Barrer $\alpha_{CO_2/N_2}=$ 48	$P_{CO_2}=$ 474 Barrer $\alpha_{CO_2/N_2}=$ 56	(Quan et al. 2017)
Pebax	4	0.8 wt%	$P_{CO_2}=$ 62 Barrer $\alpha_{CO_2/N_2}=$ 62.0 $\alpha_{CO_2/CH_4}=$ 27	$P_{CO_2}=$ 64.0 Barrer $\alpha_{CO_2/N_2}=$ 90.3 $\alpha_{CO_2/CH_4}=$ 34	(Dai et al. 2016)
PEBA	3	0.1 wt%	$P_{CO_2}=$ N/A $\alpha_{CO_2/N_2}=$ N/A	$P_{CO_2}=$ 100 Barrer $\alpha_{CO_2/N_2}=$ 91	(Shen et al. 2015)
SPEEK	1.5	8 wt%	$P_{CO_2}=$ 15.5 Barrer $\alpha_{CO_2/CH_4}=$ 26.7	$P_{CO_2}=$ 13.9 Barrer $\alpha_{CO_2/CH_4}=$ 31.5	(Xin et al. 2015)

The synthesis of GO is more straightforward compared to graphene whereby GO nanosheets are merely the oxidized form of graphite flakes, as shown in Figure 2.3. Since graphene sheets easily agglomerate or restack due to high aspect ratio and strong interparticle van der Waals interaction, the oxidation process must be performed on graphene sheets to form GO. This process introduces new oxygen-based functionalities on graphene which eventually reduces the interparticle interactions and avoids the restacking of GO (Konios et al. 2014). GO nanosheets have edges and basal planes which consist of epoxide, hydroxyl, and carboxylic acid groups. The epoxy and hydroxyl groups are bound on the basal planes, while the carboxyl groups are attached to the edges of the basal planes (Park and Ruoff 2009). These oxygen functionalities enable GO to be readily dispersed and have stronger bonding with polymer chain during membrane development process (Tanaka and Iijima 2014).

In recent years, GO has seen the widespread application as one of the most prominent fillers in water purification membrane development

(Ganesh, Isloor, and Ismail 2013, Hegab and Zou 2015, Mahmoud et al. 2015, Zinadini et al. 2014). Excessive studies have been carried out in this field due to its unique one-atom-thick structure which exhibits a larger specific surface area, excellent mechanical properties, and high dispersibility properties. However, very limited works have been reported on the utilization of GO in gas separation MMM development, as summarized in Table 2.3. From Table 2.3, it was observed that GO nanosheets have extensive influence on the gas transport process, especially on the polar/non-polar gas separation application. Generally, polymeric membrane suffers from low selectivity in binary gas separation. Thus, the addition of GO serves to enhance the selectivity in which horizontally oriented GO nanosheets could act as barriers in the polymer matrix that impedes the diffusion pathway of large gas molecules while allowing small gas molecules to permeate (Li, Ma, et al. 2015b). The sheet-like structure of GO also generates a rigidified polymer-filler interface which favors the diffusion of small molecules with minimum resistance (Li, Cheng, et al. 2015).

Dai et al. developed ultrathin PEBAX-GO MMM and performed gas permeation test using N_2, CH_4, H_2, and CO_2 gases (Dai et al. 2016). The gas permeation test revealed that CO_2 has the highest permeability across the membrane followed by H_2, CH_4, and N_2. This resulted from the high solubility of CO_2 in GO-based membrane due to the presence of polar groups especially carboxyl (-COO^-) and hydroxyl (-OH) on GO which facilitates the transport of CO_2 compared to other gases. Meanwhile, it was also found that the permeability of gases decreases drastically with increasing GO loading in the polymer matrix. GO is believed to be impermeable and its presence in the polymer matrix impedes the direct diffusion of gases, which resulted in lower permeability. On the other hand, the selectivity of CO_2/N_2, CO_2/H_2, and CO_2/CH_4 seems to be increasing with increased GO loading. Thus, it is important to determine the optimum loading of GO which could enhance the gas selectivity without sacrificing the permeability in binary gas separation. In another work conducted by Shen et al. no defects were observed on the GO-PEBAX MMM, even after continuous permeation of CO_2 and N_2 for 6000 min (Shen et al. 2015). The performance of the membrane was also consistent over the entire test duration, which corroborates the stability of GO-based MMM for practical application.

Furthermore, the oxygen-based polar functionalities on GO provide a strong GO-polymer interaction (Pei and Cheng 2012). The strong covalent grafting of polymer chains on the GO component enhances the polymer-GO interaction and hinders the formation of undesirable interfacial defects on the membrane surface. On top of that, GO is also found to be an amphiphilic material in which the surface contains both hydrophobic and hydrophilic domains (Kim et al. 2010). This property enabled GO to be readily bonded with either polar or non-polar polymer. Hence, with the aforementioned properties, GO appears to be an attractive candidate for highly selective MMM development.

3. POLYHEDRAL OLIGOMERIC SILSESQUIOXANE

Polyhedral Oligomeric Silsesquioxane (POSS) is a new class of organic-inorganic molecule which is made up of a rigid, crystalline silicon and oxygen inner inorganic framework and an outer organic framework that consists of an array of organic and inorganic substituents. The general formula of POSS is $[RSiO_{3/2}]_n$ with n ranging between 6-12 depending on the cage structure. R can be hydrogen, alkyl, olefin, alcohol, acid, amine, epoxy sulfonate group and others (Gnanasekaran 2016). The general chemical structure of POSS is shown in Figure 3.1. (d). POSS has a diameter that ranges between 1-3 nm and a core diameter of 0.5 nm (Kuo and Chang 2011). Due to its size, it can be considered as the smallest particle of silica or more commonly known as molecular silica (Li et al. 2001). Unlike silica or silicones, it is set apart by its organic outer corner which allows it to be more compatible with polymers and biopolymers.

POSS encompasses various molecular architecture which can be categorized into non-cage and cage structure. The non-cage structure can be further classified into partial cage structure, ladder structure, and random structure. Cage structure comes in T_8 cage, T_{10} cage and T_{12} cage as shown in Figure 3.1. (d), (e) and (f). Prior to the early 2000s, researches were more focused on the random and ladder structured POSS. These include poly(phenylsilsesquioxane), poly(methylsilsesquioxane) and poly-(hydrido-silsesquioxane) (Baney et al. 1995). Recently, the focus has been placed on POSS with specific cage structure.

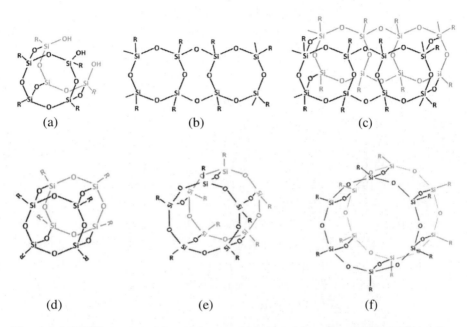

(a) (b) (c)

(d) (e) (f)

Figure 3.1. POSS structure (a) partial cage, (b) ladder, (c) random, (d) T_8, (e) T_{10} and (f) T_{12} (Kuo and Chang 2011).

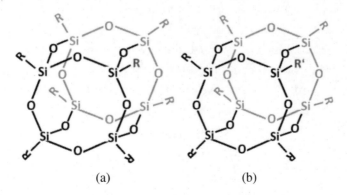

(a) (b)

Figure 3.2. POSS Structure (a) T_8R_8 and (b) T_8R_7R'.

The truly unique feature of POSS lies in its outer organic framework, R and its ability to be functionalized into the various functional group, rendering it reactive or non-reactive. The different functional group of R contributes to the development of different POSS such as molecular silica, mono-functional POSS, and multi-functional POSS. Molecular silica has a non-reactive functional group on all 8 vertexes where $R = R$; mono-functional POSS has one reactive functional group where $R' \neq R$ and multifunctional POSS has a reactive functional group on all 8 vertexes where $R = R$ (Ayandele, Sarkar, and Alexandridis 2012). This allows POSS to have various types of interactions with other materials (Kuo and Chang 2011). In addition to that, POSS derivatives are also odorless, non-volatile and environmentally friendly. The general formula that represents molecular silica and multifunctional POSS is T_8R_8, whilst mono-functional POSS is represented by the general formula T_8R_7R,' as shown in Figure 3.2.

In 1946, completely condensed or caged POSS was first synthesized by Scott via the isolation of organic compounds obtained through thermolysis of methyltrichlorosilane and dimethylchlorosilane (Scott 1946). 20 years later, partially condensed POSS is developed via the hydrolysis of cyclohexyltrichlorosilane by Brown and Vogt (Brown and Vogt 1965). In 1991, a large-scale research was funded by the Air Force Office of Scientific Research (AFOSR) to study the potential of POSS in producing lighter, stronger and more thermally resistant aerospace materials (Phillips, Haddad, and Tomczak 2004). Currently, commercially available T_8 structured POSS bears functional groups such as acrylates, alcohols, amines, carboxylic acids, methacrylates, epoxides, halides, silanes, and olefins. Regardless of their functional group and structure, the common property shared by these POSS includes the ability to improve mechanical strength, thermochemical performance, material toughness, resistance to flame and heat, hydrophobicity, adhesion, dispersion and resistance to material aging (Gnanasekaran 2016, Ayandele, Sarkar, and Alexandridis 2012). Given the properties of POSS, the resultant polymer or nanocomposites usually experience enhancement in terms of its mechanical, chemical and thermal stability whilst simultaneously retains its lightweight and ductile features. A wider range of applications is therefore available with these enhanced nanocomposites in elastomers, thermoset, thermoplastics, materials for drug delivery, dental and medical applications, materials which possess low dielectric, fire retardant, laser, space resistant, weather resistant, scratch and wear resistant properties, fuel cells, tire, and others.

3.1. Application of POSS in Gas Separation Membrane

Unlike most conventional fillers, an initiative to incorporate POSS into membranes in the field of gas separation is rather new. Generally, POSS can be incorporated via chemical cross-linking or physical blending. The properties of the resultant POSS-based MMM depend greatly on their incorporation strategy as the interactions between POSS and the polymer matrix are mediated by the ligands surrounding POSS. The bonding between POSS ligand and the neighboring polymer influences the spatial distribution of inorganic filler within the polymer matrix thus alters the thermal stability, crystallinity, morphology, and gas separation performance of the MMMs. A comprehensive summary on POSS-based MMMs and their gas separation performance has been summarized in Table 3.1. In the subsequent sections, the effect of each incorporation technique will be discussed in detail.

Table 3.1. Gas separation performance of POSS-based MMM

Polymer	Type of POSS	Findings	Ref
Polystyrene (PS) Crosslinked membrane	Styrylisobut yl-POSS	1. Excellent POSS distribution below 20 wt%. 2. Decreased T_g at loading below 20 wt% compared to PS. 3. Increased T_g at 20 wt% loading compared to PS. 4. O_2 gas permeation increased with POSS loadings. 5. Selectivity of O_2/N_2 decreased with increased POSS loadings.	(Ríos-Dominguez et al. 2006)

Table 3.1 (Continued)

Polymer	Type of POSS	Findings	Ref
Polydimethylsiloxane-Polyurethane (PDMS-PU) Crosslinked membrane	1. Partially caged hepta-cyclopentyltricyclohep-tasiloxane (CyPOSS) 2. Octakis-(hydrofomethylsiloxy) octasilsesquioxane (POSS-H)	1. Homogenous morphology in Cy-POSS/PDMS-PU MMM. 2. Addition of POSS-H in Cy-POSS/PDMS-PU MMM led to the formation of aggregates. 3. T_g of Cy-POSS MMM is decreased when POSS-H is added. 4. T_g increased with Cy-POSS and POSS-H loadings. 5. CO_2 and O_2 gas permeation increased when POSS-H is added. 6. O_2/N_2 and CO_2/N_2 selectivity decreased with the addition of POSS-H.	(Madhavan and Reddy 2009)
Polyether-b-amide (Pebax) MH 1657) Crosslinked membrane	1. Octa(3-hydroxy-3-methylbutyl dimethylsiloxy)POSS (POSS-OH) 2. Octa amic acid POSS (POSS-acid)	1. Excellent molecular level dispersion of POSS-OH and POSS-acid in Pebax matrix. 2. T_g decreased upon addition of POSS-OH up to 20 wt%. 3. T_g decreased upon addition of POSS-acid up to 5wt%. 4. T_g increased at POSS-acid 20 wt% compared to Pebax. 5. CO_2 gas permeation and CO_2/H_2 selectivity increased with the addition of POSS-H and POSS-acid up to 2 wt%.	(Li and Chung 2010)
Polymer Intrinsic Microporosity (PIM-1) Physically blended membrane	Octa-phenethyl POSS (Phenethyl POSS)	1. Homogenous distribution of PhE-POSS up to 30 wt%. 2. Phase separation occurred in MMMs above 30 wt% POSS loadings. 3. CO_2, N_2, O_2 and CH_4 gas permeability increased in MMMs. 4. Plasticization effect of CO_2 is observed with increasing pressure. 5. CO_2/CH_4 gas pair selectivity showed improvements in 1 wt% PhE-POSS/PIMs MMM.	(Konnertz et al. 2017)
Polymers with Intrinsic Microporosity (PIM-1) Physically blended membrane	Poly(ethylene glycol) functionalized POSS (PEG-POSS)	1. Homogeneous dispersion of PEG-POSS across MMM with no visible agglomerations up to 20 wt%. 2. Thermal stability decreased upon addition of POSS. 3. CO_2 gas permeability decreased with increased PEG-POSS loadings. 4. CO_2/CH_4 selectivity increased by 150% compared to pristine PIMs.	(Yang et al. 2017)

3.1.1. Chemically-Crosslinked POSS MMM

One of the earliest POSS MMM was developed in 2006 by Rios-Dominguez, Ruiz-Trevno et al. in which POSS was used as a copolymer to polystyrene membrane. Stytrl-isobutyl POSS and tetramethylpiperidinil-1-oxy macroinitiator was synthesized via solution polymerization and was further polymerized with polystyrene (PS) to form a dense membrane. Homogeneous distribution of POSS was achieved with 1wt% and 5wt% POSS loading. However, nano-sized agglomerates were formed and phase separation occurred at 10wt% and 20wt% POSS loading, respectively. The glass transition temperature (T_g) measured for the PS films synthesized with 1, 5 and 10 wt% decreased in comparison to pure PS. On the contrary, T_g of POSS MMM at 20wt% was higher than pure PS (Ríos-Dominguez et al. 2006). The phenomena observed were attributed to the loadings of POSS incorporated, as a small amount of POSS may lead to flexible polymer chains. However, as the amount increases, it may start to hinder the large-scale segmental motion of the polymer chains. The permeability of O_2 was found to increase with POSS particle loadings. This was attributed to the increment in fractional free volume caused by the incorporation of bulky POSS. Besides, the enhancement of O_2 permeability was contributed by the increased interactions between O_2 gas and the high concentration of silicon and oxygen atoms that constitute the POSS cage. Unfortunately, the increment of O_2 permeability did not lead to enhancement in O_2/N_2 selectivity.

The addition of partially caged hepta-cyclopentyltricycloheptasiloxane POSS (Cy-POSS) and polydimethylsiloxane-polyurethane (PDMS-PU) via chemical cross-linking exhibited a homogenous membrane morphology. On the contrary, the subsequent addition of octakis-(hydrofomethylsiloxy) octasilsesquioxane (POSS-H) has led to the formation of heterogeneous membrane morphology. This is most likely due to the incompatible nature between POSS and polymer matrix. Thus, chemical-crosslinking of polymer and POSS does not restrict the formation of POSS aggregates (Madhavan and Reddy 2009). The morphology of the membranes also affects their T_g

wherein PDMS-PU/Cy-POSS membrane with a homogenous morphology possesses a higher T_g compared to PDMS-PU/Cy-POSS/POSS-H membrane. This is due to the incompatible nature of POSS-H and PDMS which has led to the formation of free volume. The free volume acts as a spacer within the polymer chain, interrupting the formation of hydrogen bond between polymer matrices which has contributed to the overall depression of T_g in the MMMs. The addition of bulky and incompatible POSS has also led to the decrement of MMMs' T_g via the formation of free volume in the system. Incorporation of Cy-POSS in PDMS-PU has led to the decrement of O_2, N_2 and CO_2 gas permeability whilst the incorporation of POSS-H in PDMS-PU increased the gas permeability. The increment in gas permeability is attributed to the formation of free volume caused by the disruption of the polymer chain due to the incompatibility between POSS-H and PDMS-PU polymer matrix. On the other hand, the dispersed Cy-POSS may have acted as a blocking agent producing a highly tortuous route for penetrant gas, thus, decreasing its permeability. Hence, compatibility of POSS and polymer matrix has the ability to alter the permeability of penetrant gas.

Successful incorporation of octa(3-hydroxy-3-methylbutyldi-methylsiloxy) POSS (POSS-OH) of up to 20wt% in Pebax MH1657 polyether-b-amide membrane was carried out by Li and Chung (2010). The T_g exhibited by POSS-OH MMMs experienced drastic changes with POSS-OH loading albeit the membranes' homogeneous morphology. At low concentration, POSS-OH is responsible for the increment in accessible free volume in the polymer matrix leading to the decrement in T_g. As the loading increased, the increased hydrogen bonding between POSS-OH and polymer matrix may constraint the mobility of the polymer chain causing significant chain rigidification that increased the T_g. These changes observed in the MMMs of various POSS-OH loadings has led to the increment and decrement pattern experienced by CO_2 permeability. In addition to that, it was found that the higher free volume induced by low POSS-OH concentration has contributed to the increase in CO_2 permeability and subsequently an improvement in CO_2/H_2 selectivity of the membrane.

3.1.2. Physically Blended POSS MMM

In another work, octaphenyl POSS (OPS) and octaaminophenyl POSS (OAPS) were physically blended into polyimide (6FDA-MDA) matrix (Iyer and Coleman 2008). The incorporation of OPS led to the formation of phase separation while OAPS showed similar morphology with pure 6FDA-MDA. The presence of amine functional group on the vertex of POSS prompted stronger intermolecular hydrogen bonding between POSS and the polymer matrix whilst the phenyl group on OPS inhibited the formation of hydrogen bonding. Upon further investigation, the amount of OAPS added into the polymer matrix also dictates the morphology of the MMM. At 50 wt% OAPS loading, distinct phase separation with the formation of micron-sized POSS aggregates was observed. In a physically blended system, a strong interaction with a rigid inorganic filler will normally lead to the increment of T_g whilst a weak interaction has an opposite effect. Thus, similar to the morphological differences observed between OPS and OAPS MMMs, the latter had experienced an enhancement of T_g. However, at 50wt% OAPS, a decrement of T_g was experienced mainly because of the formation of phase separation. This has led to the formation of two polymer segments within the MMMs which consists of entrapped POSS aggregates and the bulk polymer. The polymer chain of the bulk polymer is anticipated to be less rigid than those entrapped in POSS, thus increasing its mobility. Pure helium (He), O_2, N_2 CH_4 and CO_2 gas permeability were measured for 0wt%-20wt% OAPS MMMs. A decreasing trend in permeability was observed for all gases except CO_2 with increasing loading. The increment of CO_2 gas permeability is ascribed to the presence of an amine group on OAPS which increased the solubility of acid gas such as CO_2. Nonetheless, the increment in selectivity of O_2/N_2, He/CH_4, and CO_2/CH_4 was observed. This is due to the larger decrement in permeability of larger penetrant gas. The increment of gas pair selectivity is caused by the reduction of free fractional volume contributed by the space filling effect of OAPS and the rigidification of the polymer chain by the strong interaction between OAPS and polymer matrix.

In a more recent study, poly(ethylene glycol) functionalized POSS (PEG-POSS) was physically blended into the polymer with intrinsic microporosity (PIM-1). A homogeneous dispersion of PEG-POSS was observed up to particle loading of 20wt%. Gas permeation properties of PIM-1 and MMMs were investigated with N_2, CH_4, and CO_2. It showed that the permeability of gases had decreased with increasing PEG-POSS loading. The decrement in permeability is determined to be the result of the space-filling effect of POSS and chain rigidification. In addition to that, it was also determined that the kinetic diameter of CO_2 (0.33nm) CH_4 (0.38nm) and N_2 (0.364nm) were larger than the pore size of PEG-POSS (0.32nm) thus making it impermeable to the gases. This created a tortuous pathway for the gas to permeate therefore decreasing its permeability. However, like the aforementioned study, the increment in selectivity was observed for CO_2/N_2 and CO_2/CH_4 gas pair.

3.2. Factors Affecting POSS Distribution and MMM Properties

Generally, a more homogeneous dispersion of POSS in MMMs will be observed when it possesses a more reactive functional group and is cross-linked into the polymer matrix in comparison to physically blended POSS MMMs. The degree of dispersion is seen to be most homogeneous in the order of multi-functional POSS to mono-functional POSS in a cross-linked system (Cordes, Lickiss, and Rataboul 2010). On the contrary, physical blending method often leads to macroscopic separation of POSS and polymer matrix due to the absence of a covalent bond or intermolecular force. The absence of covalent bond indicates that the interaction between POSS and the polymer matrix is governed by intramolecular interactions such as Van der Waals forces, hydrogen bonding or ion-dipole forces. These are highly dependent on the surface interaction and polarity of POSS (Striolo, McCabe, and Cummings 2005). The strong interparticle interactions between POSS (i.e., Van der Waals forces) often lead to its agglomeration in the polymer matrix. Moreover, the distribution of POSS in the polymer matrix is also dependent on its solubility in the solvent selected for physical blending. However, through proper selection of POSS, a homogeneous distribution of POSS via physical blending can be achieved.

In essence, a homogeneous distribution of POSS will lead to enhancement in MMMs thermal stability. The thermal stability increased according to the size of an inert group located at the outer corner of POSS (Tanaka, Adachi, and Chujo 2009). Since no covalent bonds are formed between POSS and the polymer, the thermal stability of MMMs is highly dependent on the intrinsic thermal stability of native POSS. In contrary, reports of diminished thermal stability in chemically cross-linked POSS MMMs are rarely observed as a nanoscale dispersion of POSS is usually achieved in addition to the formation of strong tethered or cross-linked structure. The decrement of thermal stability in chemically cross-linked POSS MMMs is usually due to the incompatible nature of POSS with the polymer. This suggests that it is essential to evaluate the compatibility between POSS and polymer for thermal stability enhancement.

The factors contributing to the enhancement or decrement of glass transition temperature (T_g) of POSS MMMs varied for each method. The common phenomenon observed in physically blended POSS MMMs is the presence of POSS aggregates. In this case, the size of POSS aggregates plays a major role in determining the T_g of POSS MMMs. Presence of coarse aggregates may increase T_g value as its size restricts the mobility of the polymer chains. On the other hand, finer aggregates may be small enough to diffuse between the polymer chains acting as a molecular lubricant increasing polymer chain mobility (Raftopoulos and Pielichowski 2016, Sánchez-Soto, Schiraldi, and Illescas 2009, He et al. 2009). Moreover, the T_g of chemically cross-linked POSS MMMs is dependent on factors such as density of cross-linking and the amount of POSS. Generally, a high cross-linking density between the polymer matrix and POSS leads to a higher restriction of chain mobility. Thus, depending on the amount of reactive vertex group present on POSS, enhancement of T_g can be achieved (Bian and Mijović 2009). The interaction between POSS and polymer matrix in a cross-linked system is significantly affected by the amount of POSS incorporated. Due to limited free volume available to accommodate POSS between the polymer matrix, aggregations of POSS may occur at higher loadings. Similar to the observation made in the physically blended system, depending on the size of these aggregates, POSS may either act as a plasticizer or mobility inhibitor (Milliman, Ishida, and Schiraldi 2012). Despite this, when incompatibility is demonstrated by POSS MMMs, the repulsive force between POSS and polymer matrix may lead to the development of free volume around the POSS aggregates. These free volumes facilitate the displacement of polymer chains, resulting in their mobility and decreased T_g.

4. CLAY MINERALS

Clay minerals are fine-grained, natural material which consists of sheet-like geometry. They exist in nature as tactoids with hundreds to thousands of silicate layers wherein each layer is generally referred to as phyllosilicate. The elemental structure of clay minerals consists of silica tetrahedral and alumina octahedral sheets with a different arrangement (Mittal 2009). For instance, 1:1 clay minerals contain one tetrahedral and one octahedral sheet; in 2:1 clay minerals, one octahedral sheet is sandwiched between two tetrahedral sheets, and 2:1:1 clay minerals are composed of an octahedral sheet adjacent to a 2:1 layer (Jamil, Ching, and Shariff 2016). Figure 4.1 shows the crystal lattice of 2:1 phyllosilicate. Each layer is approximately 1 nm thick and lateral dimension varies from 300 Å to several micrometers. The aspect ratio is usually greater than 1000 (Anadão et al. 2014).

The presence of charge in the tetrahedral and octahedral sheets influences the layered structure of clay minerals. The electronegative nature of the silicate layers attracts the exchangeable cations (like Li^+, Na^+, Rb^+, and Cs^+) in interlayer gallery spacing (Chen-Yang et al. 2007, Zulhairun et al. 2014). The replacement of an element with another element in mineral crystal without modifying its chemical structure is known as isomorphous substitution and mainly results in charge development. For example, Al^{+3} can replace Si^{+4} in tetrahedral coordination, and replacement of Al^{+3} is possible by Mg^{+2}, Fe^{+2}, and Fe^{+3} in octahedral coordination (Uddin 2008). In montmorillonite (MMT) clay, divalent Mg^{+2} replaces Al^{+3} and this creates surface charge disturbance which is balanced by Na^{+1} or Ca^{+2} ions. The interlayer spacing varies according to the size of the ions. Since these ions have an affinity for polar groups, water, and other polar solvents can easily migrate inside the layer and cause it to expand (Pavlidou and Papaspyrides 2008). This causes MMT to possess high cationic exchange capacity, this is a promising inorganic filler for MMM development (Hashemifard, Ismail, and Matsuura 2011).

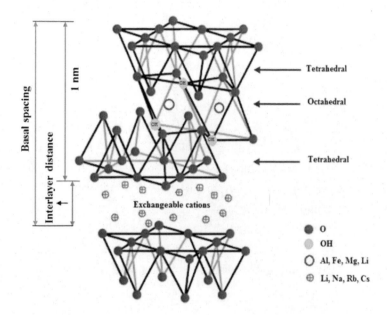

Figure 4.1. Structure of 2:1 phyllosilicate.

4.1. Classification of Clay Minerals

Clay minerals are further classified into subgroups such as smectite, illite, and kaolinite, as shown in Table 4.1. The kaolinite group consists of three members and each has a formula of $Al_2Si_2O_5(OH)_4$. The same molecular formula indicates that the members of this group are polymorphs, which means that they possessed the same molecular formula but different structures. Each member contains silicate sheets (Si_2O_5) bonded to aluminium hydroxide layer $Al_2(OH_4)$ (Uddin 2008). These types of clays are abundantly used in ceramics, paint, rubber, and plastics.

Montmorillonite, a member of the smectite family, has been extensively used as reinforcing filler in the automotive industry since the 1990s. Toyota Research group emerged as the pioneer in this regard, successfully commercializing nylon-6-MMT nanocomposite (Nguyen and Baird 2006). According to their findings, with a small addition of MMT (4.2 wt%), the elastic modulus and tensile strength increased by 50%. Since then, researchers have thoroughly investigated MMT application in thermoplastic nanocomposites due to their superior mechanical, barrier, thermal, flame retardant and abrasive properties (Okada and Usuki 1995, Usuki et al. 1995, Kojima et al. 1995). Along with MMT, bentonite, hectorite, saponite and laponite are clays that are used as reinforcing fillers in polymer composite industry (Herring 2006). These clay fillers play an important role in the polymer nanocomposite industry due to their economic viability, availability, flame retardancy and reinforcement characteristics. The interlayer space or "gallery," intercalation ability and exfoliation of these layered clays enhanced their attractiveness in polymer nanocomposites (Okada and Usuki 1995, Ahmadi, Huang, and Li 2004).

Table 4.1. General types of clay (Uddin 2008)

SN	Group Name	Member Minerals	General Formula
1	Kaolinite	Kaolinite, dickite, nacrite	$Al_2Si_2O_5(OH)_4$
2	Smectite	Montmorillonite, pyrophyllite, talc, saponitenontonite	$(Ca,Na,H)(Al,Mg,Fe,Zn)_2(Si,Al)_4O_{10}(OH)_2\text{-}xH_2O$
3	Illite	Illite	$(K,H)Al_2(Si,Al)_4O_{10}(OH)_2\text{-}xH_2O$

4.2. Surface Modification of Clay Platelets

Despite the many attractive properties of layered silicates, their hydrophilic nature limits their suitability to be mixed and interact with most polymer matrices. Moreover, the stacks of clay platelets are held together by electrostatic forces, whereby the counterions can be shared by two neighboring platelets, resulting in stacks of platelets that are held tightly together. Nanocomposites employing untreated clay would not demonstrate much effectiveness, because most of the clay would form aggregates, involving very limited interaction between the matrix and the individual platelets (Jamil, Ching, and Shariff 2016).

Generally, two types of chemical modification (non-covalent or covalent) can be performed on layered silicates. The non-covalent modification involves intercalation modification and ion exchange without covalent bonding through hydrogen bonding, Van der Waals interaction, dipole-dipole interactions and acid-base reactions (Annabi-Bergaya 2008). Ion exchange is a relatively rapid method to achieve organophilicity of clay. For the ion exchange method, the interlayer ability to swell plays an important role. If alkali is present between the layers of clay, swelling is possible because divalent or trivalent atoms hinder the water molecules from penetrating the layers and inhibit the swelling process. The swelling flakes increase the selectivity for small molecules by creating a more tortuous path for larger molecules (Defontaine et al. 2010). The common types of alkali cations used are sodium-based alkali cations such as octosilicate $(Na_8[Si_{32}O_{64}(OH)_8].32H_2O$ also known as ilerite or RUB-18, α-Na_2 Si_2O_5, and β-$Na_2Si_2O_5$, as well as non-sodium based alkali cations such as $KHSi_2O_5$, and $LiNaSi_2O_5.2H_2O$ (silinaite) (Jamil, Ching, and Shariff 2016). The alkali cations present between the clay layers provides the opportunity for organic cations and surfactants to replace them. Furthermore, the presence of interlayer organic cations decreases the surface energy, thus, the interaction between the polymer and modified clay improved significantly. The long-chained surfactants tethered at the surface of clay resulted in increased gallery space. This will attract polymer chains to diffuse into the gallery space and improve interaction towards polymer matrix. Ultimately, the compatibility of filler towards polymer is also tremendously improved (Kim, Choi, and Nair 2011). Figure 4.2 shows the ion exchange reaction of the layered silicate, in which Na^+ ion is replaced by voluminous onium ion. After surface modification of the layered silicate with surfactant molecules, the interlayer distance increased.

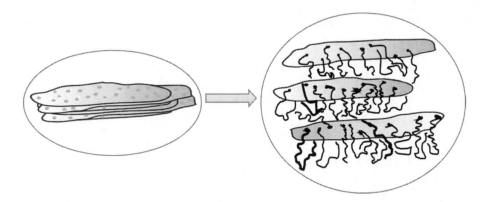

Figure 4.2. Schematic picture of ion exchange reaction. The relatively small inorganic ions (i.e., Na^{+1}) are exchanged by more voluminous organic onium cations.

Table 4.2. Types of commercially available montmorillonite and their properties (Jamil, Ching, and Shariff 2016)

Filler	Interlayer cations	Interlayer distance (Å)
Cloisite Na	Na^+	12.1
Cloisite 10A	$(CH_3)_2N^+CH_2(C_6H_6)$(Hydrogenated Tallow)	19.2
Cloisite 20A	$(CH_3)_2N^+$(Hydrogenated Tallow)$_2$	22.1
Cloisite 25A	$(CH_3)_2N^+$(Hydrogenated Tallow)(2-ethylhexyl)	20.7
Cloisite 30B	$(CH_3)_2N^+$(Tallow)$(CH_2CH_2OH)_2$	18.5
Nanomer 1.28E	$CH_3N^+(CH_2)_{17}CH_3$	24.0-26.0
Nanomer 1.30E	$H_3N^+(CH_2)_{17}CH_3$	18.0-22.0
Nanomer 1.31PS	$H_3N^+(CH_2)_{17}CH_3,H_3N^+(CH_2)_3Si(OC_2H_5)_3$	18.0-22.0
Nanomer 1.34TCN	$CH_3N^+(C_2H_4OH)_2$(Hydrogenated Tallow)	18.0-22.0
Nanomer 1.44P	$CH_3(CH_2)_{17}N^+(CH_3)2(CH_2)_{17}CH_3$	24.0-26.0
Nanofil 757	Na^+	12.2
Nanofil 15	$(CH_3)_2N^+$(Hydrogenated Tallow)$_2$	29.0
Nanofil 919	$(CH_3)_2N^+$(Tallow)$(CH_2C_6H_5)$	18.8
Nanofil 804	CH_3N^+(Tallow)$(OH)_2$	18.0

On the other hand, covalent bonding occurs through silylation, condensation, and esterification of $SiOH/SiO_2$ groups. The bonding forces involved in non-covalent modifications provide weaker interactions than covalent modifications. Besides that, organic cations provide functional groups that can interact with or initiate polymerization of monomers to improve the strength of interface adhesion between the inorganic component and polymer phase (Takahashi and Kuroda 2011). Thus, covalent modification not only increases the filler distribution but improves the MMM performance as well. Table 4.2 summarizes some industrially available modified MMT and their interlayer distances.

4.3. Morphology of Polymer-Clay MMM

The most crucial factor when fabricating enhanced performance MMMs for gas separation is the material selection. In MMMs, polymer phase enhances permeation whereas dispersed fillers assist in improving selectivity. When layered silicates are incorporated, three different types of morphological changes may take place; phase separated, intercalated or exfoliated, as shown in Figure 4.3 (Goh et al. 2011b). These morphologies offer significant improvement in terms of mechanical, thermal and barrier properties of polymer-nanoclay hybrid materials (Zulhairun et al. 2014).

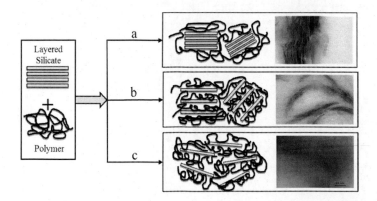

Figure 4.3. Schematic diagrams and TEM micrographs showing a) phase separated, b) intercalated and c) exfoliated clay-polymer interfacial morphology.

Principally, layered silicate affects the sorption of condensable gas by diffusion pathways obstruction and reduction of free volume in polymer systems (Silvestre, Duraccio, and Cimmino 2011, Villaluenga et al. 2007, De Azeredo 2009). The exfoliated morphology creates a tortuous path for penetrating gas molecules, as a result, selectivity increases at the expense of reduced permeance. Intercalation gives enhanced permeation but the risk of surface defects and void formation remains high due to lower filler distribution. The interfacial morphology between polymer and filler is also crucial in defect-free membrane development. The interface interaction decides the selective passage of one gas over the other, thus affecting the permeability and selectivity of gas molecules (Rezakazemi et al. 2014a). As a result, poor interaction leads towards non-ideal morphologies.

4.4. Limitations Associated with Clay Minerals for MMM Development

It has been noted that various challenges are encountered by researchers in successful membrane formation. Membrane fabrication is virtually influenced by many variables, which in turn will affect the membrane's physical properties and gas separation performance This section presents the problems and proposed solutions associated with successful membrane development.

4.4.1. Clay Agglomeration in Polymer Matrix

The agglomeration of nano-sized clay particles in the polymer matrix is the foremost challenge in MMM development. Generally, nano-clays disperse poorly in the polymer matrix, thus are likely to agglomerate in the MMM (Mansur 2014). Appropriate clay dispersion increases the tortuosity and diffusion pathways for the gas molecules while clay agglomeration provides low resistance pathways for the gas transport by decreasing the aspect ratio of the nano-clay (Jamil, Ching, and Shariff 2014). In addition, the incompatible properties of clay and polymer phases lead to filler agglomeration. It has been observed by various researchers that when polar clay particles were dispersed in apolar polymers, the filler tend to agglomerate and deteriorate the gas separation properties of the developed MMMs (Behroozi and Pakizeh 2017). Jamil et al. incorporated montmorillonite in polyetherimide and observed filler agglomeration with increased loading (Jamil, Ching, and Shariff 2018). Consequently, gas separation performance of the developed membranes showed a decreasing trend with MMT loading compared to neat PEI membrane.

Furthermore, clay agglomeration leads to the formation of numerous stress concentrated points which can deteriorate the mechanical stability of MMM, especially at high inorganic filler concentrations. Incorporation of cloisite 15A in Pebax showed decreased tensile strength and elongation at break due to the agglomeration of filler at higher loading (Behroozi and Pakizeh 2017). It was observed that the stronger filler-filler interaction formed agglomeration that reduced the tensile strength of the membranes due to local stress concentration. The mechanical strength of the membrane refers to the efficacy of stress transfer between the polymer matrix and filler phase. Filler agglomeration results in weaker interactions between two phases hence disrupt stress transfer and deteriorate the mechanical properties (Fu et al. 2008, Jaafar et al. 2011).

4.4.2. Void Formation at Clay Surface

Poor interfacial interaction of polymer chains and filler surface cause the formation of voids which result in poor membrane performance in terms of gas separation. Voids formation at the filler surface is also termed as "sieve in a cage" or "leaky interface" morphology. These voids provide the path of least resistance, causing gas molecules to pass through at a much higher permeability as compared to neat polymer membranes. In "sieve in a cage" morphology, the selectivity of the membrane remains constant or slightly higher compared to neat polymer membrane. In contrast, if the size of the voids is larger than the penetrant gas molecules, the permeability increases significantly, however, selectivity decreases. This "leaky interface" morphology diminishes the membrane characteristics to discriminate penetrating gas molecules. The permeability enhancement due to the leaky interface can be predicted by Knudsen diffusion or sorption diffusion transport (Rezakazemi et al. 2014a).

Liang et al. developed polyethersulfone (PES) based MMM by incorporating Na-MMT as the inorganic filler (Villaluenga et al. 2007). Nevertheless, at high filler loading, Na-MMT was found to agglomerate. This phenomenon is evident as the permeability for CO_2 increased with Na-

MMT loading but beyond 10 wt%, CO_2/CH_4 selectivity was greatly reduced. The authors speculate that the presence of interfacial voids leads to the decrease in membrane selectivity since gas transport occurs via Knudsen diffusion. To eliminate defects at the bulk polymer and dispersed filler interface, priming protocol and silane coupling treatment have been proposed and reported in the literature (Jamil, Ching, and Shariff 2016, Zhang et al. 2013).

4.4.3. Dispersion Morphology of Clay Platelets

Investigations have shown that the dispersion state of clay layers in polymer matrix greatly affects membrane performance. Intercalated morphology leads to better permeation whereas exfoliated morphology favors selectivity by controlling the permeation of large-sized molecules (Zulhairun and Ismail 2014b). However, when Cloisite 15A, a type of clay, was incorporated to develop asymmetric MMM, permeability increment of 270% compared to neat Psf was observed by Zulhairun (Zulhairun et al. 2014). Interestingly, there was almost no effect on membrane selectivity, which is contrary to the concept that clay fillers enhance gas selectivity. This is because selectivity enhancement is not merely dependent on clay addition; it also depends on the state of clay dispersion in a polymer matrix (Gain et al. 2005).

Figure 4.4 (a, b) shows the TEM micrographs of MMT dispersion in PEI matrix at 2 and 4 wt% respectively. The dark lines indicate MMT tactoids or non-exfoliated layers; whereas the distant, displaced layers in lighter-grey indicate the intercalated form of MMT. Exfoliated forms exist in the absence of interactions between MMT layers. It can be observed that for 2 wt%, MMT exists largely in intercalated and exfoliated morphology. Furthermore, the width of the tactoid is very low, signifying that only several layers of MMT combined to form tactoids. In contrast, when the concentration was increased to 4 wt %, MMT formed agglomerates which exist largely in tactoid and intercalated morphology, as depicted in Figure 4.4 (b). The higher CO_2/CH_4 selectivity was observed for 2 wt% MMT due to better dispersion state in PEI matrix (Jamil, Ching, and Shariff 2018).

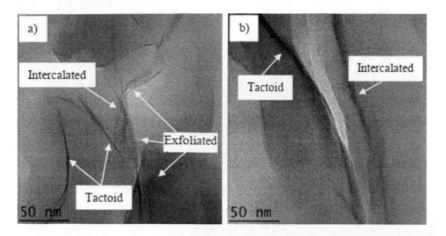

Figure 4.4. TEM micrographs of PEI-MMT MMM incorporating a) 2 wt% and b) 4 wt % MMT.

4.4.4. Polymer-Clay Incompatibility

The compatibility of nano-clay towards polymer matrix is another important aspect to be considered in mixed matrix membrane development (Jamil, Ching, and Shariff 2017). Crystalline state is impermeable in nature; thus, decreasing crystallinity of polymer phase leads to the increased permeability of penetrating gases. In previous research, highly crystalline high-density polyethylene (HDPE) and nanoclay (pristine and modified) nanocomposite films have been developed. It is observed that the crystallinity of HDPE is not affected by the addition of clay despite the permeability increase. This rise in permeability of HDPE/MMT-based nanocomposite membrane is due to the weak interaction between the polymer and filler interface. Therefore, the addition of more polar maleic anhydride as compatibilizer increases intercalation, dispersion, exfoliation, and tortuous path hence decreases permeability (Picard et al. 2007).

Moreover, during wet phase inversion method, a decrease in dope viscosity with increment in solvent/coagulant exchange capacity is observed in the presence of MMT, due to the weaker polymer-clay interaction which reduces the thermodynamic stability of polymer solution. As a result, pores are generated in the outer thin skin layers, which leads to permeation enhancement. This phenomenon was observed when hydrophilic MMT was incorporated in hydrophobic PVDF. At 5wt% MMT loading, the permeance was enhanced by 21% compared to neat PVDF (Rezaei et al. 2014).

5. TITANIUM (IV) OXIDE

Another recent focus in MMM development is to incorporate metal oxides into the polymer matrix to enhance membrane's gas separation properties. Various types of metal oxides have been used including Fe_3O_4, Al_2O_3, ZrO_2, SiO_2, and TiO_2. Among the available metal oxides, titanium (IV) oxide or titanium dioxide (TiO_2) nanoparticles have received considerable attention due to their low cost, good physical and chemical properties as well as high thermal stabilities (Liang et al. 2012, Sadeghi, Afarani, and Tarashi 2015). TiO_2 is a unique nanomaterial used in many different industries and products such as paints and coatings, nanocomposite and semiconductor. It is a nonporous spherical nanoparticle with high hydrophilicity, good chemical and thermal stabilities (Liang et al. 2012, Vatanpour et al. 2012). The usage of TiO_2 nanoparticles also helps to increase the growth rate of the membrane formation mechanism (Li et al. 2009).

The use of TiO_2 as an inorganic filler in MMM development has been reported in many articles. Generally, TiO_2-based MMM may be produced via chemical or physical approaches. The former involves cross-linking the inorganic nanoparticles with polymer matrix whilst the later pertains to the physical blending of TiO_2 nanoparticles in polymer casting solution (Vatanpour et al. 2012). Thus far, many incorporation strategies have been developed in order to improve the nanoparticles' stability and degree of dispersion in the polymer matrix. For instance, different organic functional groups were introduced chemically onto TiO_2 particle's surface to increase the compatibility of nanoparticles with the polymer matrix. Table 5.1 shows the recent applications of TiO_2 nanoparticles as an inorganic filler in MMMs for gas separation. Functionalization of TiO_2 nanoparticles' surface was also conducted in order to improve its desirable properties.

Table 5.1. Gas separation performance of TiO_2-based MMM

Polymer	Size of TiO_2 (nm)	Type of TiO_2	Findings	Ref
Matrimid 5218	3	TiO_2 (Nanoscale)	1. The free volume fraction of matrimid increased due to disruption of chain packing. 2. Pure gas permeabilities for CH_4, N_2, O_2, CO_2 and He increased with decreased selectivity when TiO_2 volume fraction increased due to nanoparticles agglomeration and void formation between TiO_2-matrimid interface.	(Moghadam et al. 2011)
Polyethersulfone (PES)	21	TiO_2 (Sigma-Aldrich)	1. Thermal stability was enhanced with TiO_2 addition. 2. Aggregation of TiO_2 was formed upon filler loading of 5wt%. 3. Highest CO_2 removal and selectivity at 5wt% of TiO_2.	(Galaleldin, Mannan, and Mukhtar 2017)
High impact polystyrene (HIPS)	10-15	TiO_2 (Nanopars Lima Corporation)	1. The permeability of N_2 was higher than CO_2 at all pressure range. 2. The permeability of N_2 and CO_2 decreased with increased feed pressure. 3. The permeability of N_2 and CO_2 increased with TiO_2 from 1wt% to 7wt% and reduced afterward.	(Safaei, Marjani, and Salimi 2016)
Pebax-1657	3.775	Silane-modified TiO_2 (AS-TiO_2) and carboxymethyl chitosan (CMC)-modified TiO_2 (CMC-TIO_2)	1. The synthesized membrane has higher thermal stability, good compatibility between polymer and nanoparticles and good dispersion of TiO_2 in the polymer matrix. 2. Pebax-CMC-TiO_2 membrane (3wt%) has the highest CO_2 permeability which demonstrated 60% increment compared to the pristine membrane. 3. Pebax-AS-TiO_2 membrane obtained highest CO_2/N_2 selectivity, which is 33% higher than the neat Pebax membrane. 4. Resulting MMMs exceeded the 2008 Robeson upper bound curve. 5. TiO_2 may be modified chemically for the enhancement of gas separation performance.	(Shamsabadi et al. 2017)

Table 5.1. (Continued)

Polymer	Size of TiO$_2$ (nm)	Type of TiO$_2$	Findings	Ref
Polyvinyl alcohol (PVA)	21	Degussa P25	1. PVA-TiO$_2$ membrane demonstrated good optical property and little agglomeration. 2. A strong interaction existed between the PVA polymer and TiO$_2$ nanoparticles. 3. Enhanced mechanical property with the addition of TiO$_2$. 4. Selectivity of gas pairs between O$_2$/N$_2$, H$_2$/N$_2$, and CO$_2$/H$_2$ increased but with decreased permeability at low TiO$_2$ loading up to 20wt%. 5. The permeability of gases was constant while selectivity of all gas pairs decreased at a TiO$_2$ loading of 20 to 30wt%. 6. At high TiO$_2$ loading ranging from 30 to 40wt%, the permeability of gases increased with decreased gas pairs selectivity.	(Ahmad, Deshmukh, and Hägg 2013)
Polysulfone (PSf)	25	Amino-functionalized TiO$_2$ using ethylenediamine (EDA)	1. Enhanced gas permeability for all gases was achieved due to F-TiO$_2$. 2. A strong interaction existed between CO$_2$ molecules and an amine group from F-TiO$_2$. 3. The permeability of CO$_2$ increased but selectivity decreased slightly at a higher weight ratio of F-TiO$_2$.	(Kiadehi et al. 2014)
Polyurethane	10-15	TiO$_2$ (TECNAN Co.)	1. Good nanoscale dispersion of TiO$_2$ within the polymer matrix. 2. For all the studied gases (N$_2$, O$_2$, CH$_4$, and CO$_2$), the permeability decreased with increase in gas selectivity as the TiO$_2$ content increased. 3. The reason is due to the increase in tortuosity of the diffusion path induced by non-porous TiO$_2$. 4. CO$_2$ gas has the highest permeability, followed by CH$_4$, O$_2$, and N$_2$.	(Sadeghi and Taheri 2014)
Polyimide (PI)	30-50	Nanofiber cellulose-modified TiO$_2$	1. An intermolecular interaction existed between PI and Cellulose/TiO$_2$.	(Ahmadizadegan 2017)

Polymer	Size of TiO$_2$ (nm)	Type of TiO$_2$	Findings	Ref
			2. Thermal stability of PI/Cellulose/TiO$_2$ increased as compared to the pristine PI membrane. 3. The permeability of CO$_2$, H$_2$, N$_2$, and CH$_4$ increased using PI/Cellulose/TiO$_2$ membrane at increasing TiO$_2$ loading.	
Cellulose acetate (CA)	5-10	Lab prepared	1. From 0-20wt% of TiO$_2$ nanoparticles used in CA, the synthesized membranes were porous and have less rough surface. Upon addition of 25wt% of TiO$_2$, membranes showed the porous structure and rough surface. 2. The permeation of CA-TiO$_2$ membrane increased with increased TiO$_2$ loading. Maximum CO$_2$/CH$_4$ selectivity was obtained with 20wt% TiO$_2$ in CA.	(Farrukh et al. 2014)
Poly(arylene ether sulfone) (PES)	60	Anatase TiO$_2$ (Sigma-Aldrich)	1. Homogeneous dispersion of TiO$_2$ nanoparticles in polymer matrix due to the amino group of PES chain, but with reduced mechanical properties. 2. Gas permeability coefficient of PES/TiO$_2$ was higher than pristine PES membrane. 3. The separation factor of synthesized membranes increased due to polar hydroxyl group interaction on the TiO$_2$ surface with polar CO$_2$ molecules.	(Yu et al. 2017)
Poly(arylene ether sulfone) (PES)	60	Anatase TiO$_2$ (Sigma-Aldrich)	4. Homogeneous dispersion of TiO$_2$ nanoparticles in polymer matrix due to the amino group of PES chain, but with reduced mechanical properties. 5. Gas permeability coefficient of PES/TiO$_2$ was higher than pristine PES membrane. 6. The separation factor of synthesized membranes increased due to polar hydroxyl group interaction on the TiO$_2$ surface with polar CO$_2$ molecules.	(Yu et al. 2017)

Table 5.1. (Continued)

Polymer	Size of TiO$_2$ (nm)	Type of TiO$_2$	Findings	Ref
Poly(arylene ether sulfone) (PES)	60	Anatase TiO$_2$ (Sigma-Aldrich)	7. Homogeneous dispersion of TiO$_2$ nanoparticles in polymer matrix due to the amino group of PES chain, but with reduced mechanical properties. 8. Gas permeability coefficient of PES/TiO$_2$ was higher than pristine PES membrane. 9. The separation factor of synthesized membranes increased due to polar hydroxyl group interaction on the TiO$_2$ surface with polar CO$_2$ molecules.	(Yu et al. 2017)
Poly(arylene ether sulfone) (PES)	60	Anatase TiO$_2$ (Sigma-Aldrich)	10. Homogeneous dispersion of TiO$_2$ nanoparticles in polymer matrix due to the amino group of PES chain, but with reduced mechanical properties. 11. Gas permeability coefficient of PES/TiO$_2$ was higher than pristine PES membrane. 12. The separation factor of synthesized membranes increased due to polar hydroxyl group interaction on the TiO$_2$ surface with polar CO$_2$ molecules.	(Yu et al. 2017)
Polysulfone (PSf)	25	TiO$_2$ (Nanoscale)	1. Homogeneous nanoparticles distribution at a lower loading of TiO$_2$ ($\leq$3wt%) but an agglomeration of TiO$_2$ was observed when TiO$_2$ loading $\geq$5wt%. 2. The surface roughness of the membrane increased. 3. At higher TiO$_2$ loading, membrane produced has higher thermal stability and a smaller amount of weight loss. 4. For both CO$_2$ and CH$_4$ gases, permeances decreased with the use of lower TiO$_2$ contents ($\leq$3wt%), while increased at higher TiO$_2$ loadings (5wt%). 5. The selectivity of CO$_2$/CH$_4$ increased at lower TiO$_2$ loading and reduced upon higher TIO$_2$ content.	(Moradihamedani et al. 2015)

Polymer	Size of TiO$_2$ (nm)	Type of TiO$_2$	Findings	Ref
PEBAX-1074	21	PEG-400 modified-TiO$_2$	1. PEG-400 and TiO$_2$ domains disrupted the interchain hydrogen bonding between the PA segments. 2. Mechanical strength and crystallinity reduced with increasing amount of PEG-400 and TiO$_2$ due to the elimination of interchain hydrogen bonding in PA. 3. PEBAX/PEG/TiO$_2$ demonstrated better gas permeation performance than neat PEBAX membrane due to effective voids formation at the interface of PEBAX/PEG and TiO$_2$.	(Azizi, Mohammadi, and Behbahani 2017)

Based on these research, the inclusion of TiO$_2$ as inorganic nanoparticle in MMMs showed improved ideal selectivity and permeability in contrast with the pure polymeric membrane. Due to the interaction between TiO$_2$ nanoparticles surface' with the polymer chain or solvent, the structure of the polymer matrix was modified, which leads to the overall improvement in the favorable gas selectivity and permeability for gas separation (Vatanpour et al. 2012). TiO$_2$ nanoparticles embedded in the polymer matrix disrupt the chain packing of polymer and increase the polymer chain free volume, ultimately increases the gas permeability (Moradihamedani et al. 2015, Shamsabadi et al. 2017). Voids formation at the nanoparticles and polymer interface also lead to the increase in membrane permeability (Moradihamedani et al. 2015).

In another work, it was reported that the improvement of membrane selectivity towards CO$_2$ gas is because TiO$_2$ nanoparticles possess high capacity for CO$_2$ adsorption compared to CH$_4$ (Matteucci et al. 2008), which increase the gas molecules' solubility in the polymer matrix. Besides that, the incorporation of TiO$_2$ in the polymer matrix was also found to increase the thermal stability of MMMs. This is due to the strong bonding of TiO$_2$ with polymer when nanoparticles act as a crosslinking point which connects the polymer chain and increases membrane rigidity (Afzan Abdullah et al. 2017).

Furthermore, TiO_2 nanoparticles possess small particle diameter (3-400 nm) and high specific area (35-500 m^2/g). This leads to improved distribution property and prevented the formation of non-selective voids at nanoparticles/polymer interface (Bastani, Esmaeili, and Asadollahi 2013). The dispersion degree of TiO_2 nanoparticles varies according to their loading in the polymer matrix. At low particle loading, the MMMs are presumed to be defect-free as the nanoparticles dispersed individually or in nanoscale aggregates. The diffusivity selectivity of gas pairs increases with increasing particle loading (Goh et al. 2011a). However, at a higher loading of TiO_2, micron-sized aggregates were formed, which indicates the presence of defects at high filler concentration. Hence, it is important to determine the optimum loading of TiO_2 nanoparticles incorporated into the polymer matrix. TiO_2 exhibits higher affinity towards certain gas molecules as demonstrated from different researches due to surface interaction. Hence, TiO_2 nanoparticles may be adopted into the design and fabrication of MMMs to facilitate and enhance gas separation (Goh et al. 2011a). However, the control of dispersion degree for TiO_2 nanoparticles lesser than 100nm in polymer matrix remains a big challenge as the surface interactions may become very strong, especially at higher nanoparticles loading (Ng et al. 2013).

6. SYNERGISTIC EFFECT OF COMBINING TWO FILLERS IN MMM FABRICATION

As previously discussed, the incorporation of the inorganic filler in the polymer matrix is expected to improve the gas separation performance as well as the physicochemical properties of the polymeric membrane (Abedini and Nezhadmoghadam 2010, Aroon et al. 2010, Chung et al. 2007, Goh et al. 2011b). MMM is developed in order to surpass the Robeson trade-off curve by providing high permeability and selectivity. Nevertheless, in most cases, the addition of a single filler could only enhance either permeability or selectivity, but not both (Rezakazemi et al. 2014b). Furthermore, the recent practice of developing MMM using nano-fillers encounters severe agglomeration issues in the polymer matrix. Generally, in order to overcome these issues, the filler particles are organo-functionalized. The newly introduced functionalities on the filler surface are expected to reduce the intermolecular interaction between filler particles, thus improve its dispersion quality in the polymer matrix. Nonetheless, the aggressive reaction conditions applied during functionalization were reported to cause significant damage and defect to the nanofiller's structure. The covalent linkage, especially at the filler surface might also disrupt the native molecular structure and physiochemical properties of the filler particles (Li et al. 2011). Above all, it is very laborious to control the degree of functionalization on the filler surface, which could make the scaling up difficult.

Therefore, instead of functionalizing the filler, a new technique of combining two different fillers with distinct properties, morphologies, nature, and dimension has been introduced (Galve et al. 2013, Valero et al. 2014, Zornoza et al. 2011). In this technique, two different fillers are added to a polymer matrix at different loading to engineer the performance of MMM. The complementary interaction between the two fillers with distinct surface chemistry could promote polymer-filler interaction, thereby reducing the possibilities of filler agglomeration in the polymer matrix, whilst retaining the native properties of the filler particles. On top of that, it is hypothesized that the addition of binary fillers could enhance both the permeability and selectivity of the polymer matrix. One of these binary fillers could improve gas permeability across the membrane while the other could enhance the selectivity. In addition, Zornoza et al. found that the presence of binary fillers could provide additional stability to achieve improved filler dispersion in the polymer matrix (Zornoza et al. 2011).

In a work conducted by Valero et al. MCM-41 and MOF were incorporated in the polyimide matrix for H_2/CH_4 gas separation (Valero et al. 2014). The mesoporosity of MCM-41 was reported to enhance the gas permeability whereby the microporosity and flexibility of MOF enhanced the selectivity of H_2 gases. The optimum gas separation performance of the resultant membrane (α_{H_2/CH_4} =178 at 21.3 Barrer of H_2 permeability) was obtained at 12/4 wt% of MCM-41/MOF filler loadings. Furthermore, the complementary interaction between these two fillers improves the dispersion quality of the filler phase and its interaction with the polymer phase without adding any compatibilizing agent. In another work reported by Zornoza et al. HKUST-1 and ZIF-8 were integrated into polysulfone to develop binary filler-based MMM for CO_2/CH_4 separation (Zornoza et al. 2011). The combined effect of adding two fillers with different properties was found to disperse the filler particles more homogeneously, thus improved the physicochemical properties of the polymer matrix together with the gas separation performance. In their work, the optimum CO_2/CH_4 separation (α_{CO_2/CH_4}= 22.4 at 8.9 Barrer of CO_2 permeability) was recorded at 8/8 wt% of HKUST-1/ZIF-8 filler loadings. Sarfaz & Shammakh also synthesized a high performance MMM using CNT and ZIF-302 binary fillers (Sarfraz and Ba-Shammakh 2016a). At 8/12 wt% of CNT/ZIF-302 loading, the resultant membrane demonstrated 177% and 218% increment in CO_2 permeability and CO_2/N_2 separation, respectively, which enabled it to surpass the Robeson upper bound limit. The resultant membrane was also reported to be thermally as well as mechanically stable and exhibited uniform filler dispersion quality with no polymer-filler interfacial defects.

Table 6.1 tabulates prior work performed using binary fillers for gas separation membrane development. Results from these works validated that the incorporation of binary fillers on the continuous polymer phase improve both permeability and selectivity of the resultant membrane, especially at its optimum loading. The binary filler-based MMM also exhibits significant improvement in the thermal stability and mechanical strength, as compared to single filler-based MMM (Tjong 2012). Therefore, this technique of adding two fillers in the polymer matrix shows very promising characteristics for gas separation membrane development.

Table 6.1. Summary of binary filler-based MMM for gas separation application

Polymer	Binary filler		Application	Polymeric Membrane	Optimum filler loading	MMM	Ref
	Filler 1	Filler 2					
Polyimide	MCM-41	MOF	H_2/CH_4	$P_{H_2}= 11.8$ Barrer $\alpha_{H_2/CH_4}=58.9$	8/8	$P_{H_2}=19.5$ Barrer $\alpha_{H_2/CH_4}=67.3$	(Valero et al. 2014)
Polysulfone	HKUST-1	ZIF-8	CO_2/CH_4	$P_{CO_2}= 4.5$ Barrer $\alpha_{CO_2/CH_4}= 22.0$	8/8	$P_{CO_2}= 8.9$ Barrer $\alpha_{CO_2/CH_4}= 22.4$	(Zornoza et al. 2011)
			CO_2/N_2	$P_{CO_2}= 5.5$ Barrer $\alpha_{CO_2/N_2}= 25.0$	8/8	$P_{CO_2}= 8.9$ Barrer $\alpha_{CO_2/N_2}= 38.0$	
Polyimide	MCM-41	JDF-L1	H_2/CH_4	$P_{H_2}= 311$ Barrer $\alpha_{H_2/CH_4}=18.9$	8/4	$P_{H_2}= 440$ Barrer $\alpha_{H_2/CH_4}=35.7$	(Galve et al. 2013)
Matrimid	Carbon Nanotube	Graphene Oxide	CO_2/CH_4	$P_{CO_2}= 8.84$ Barrer $\alpha_{CO_2/CH_4}= 34.0$	5/5	$P_{CO_2}= 38.07$ Barrer $\alpha_{CO_2/CH_4}= 84.60$	(Li, Ma, et al. 2015b)
			CO_2/N_2	$P_{CO_2}= 8.84$ Barrer $\alpha_{CO_2/N_2}= 32.74$	5/5	$P_{CO_2}= 38.07$ Barrer $\alpha_{CO_2/CH_4}= 81.00$	
Polysulfone	Graphene Oxide	ZIF-301	CO_2/N_2	$P_{CO_2}= 6.5$ Barrer $\alpha_{CO_2/N_2}= 11.0$	5/6	$P_{CO_2}= 12.5$ Barrer $\alpha_{CO_2/N_2}= 103.0$	(Sarfraz and Ba-Shammakh 2016b)
Polyimide	MWCNT	GONRs	CO_2/CH_4	N/A	N/A	$P_{CO_2}= +128\%$ $\alpha_{CO_2/CH_4}= + 125\%$	(Xue et al. 2017)
Polysulfone	Carbon Nanotube	ZIF-302	CO_2/N_2	$P_{CO_2}= 6.5$ Barrer $\alpha_{CO_2/N_2}= 11.0$	8/12	$P_{CO_2}= 18$ Barrer $\alpha_{CO_2/N_2}= 35.0$	(Sarfraz and Ba-Shammakh 2016a)
Polyimide	Titanium dioxide	Graphene Oxide	CO_2/N_2	$P_{CO_2}= N/A$ $\alpha_{CO_2/N_2}= 6.00$	N/A	$P_{CO_2}= 287.56$ $\alpha_{CO_2/N_2}= 51.81$	(Wang et al. 2017)

CONCLUSION AND FUTURE DIRECTION

In conclusion, alternative fillers such as carbonaceous nanofillers, polyhedral oligomeric silsesquioxane (POSS), clay minerals and titanium dioxide (TiO_2) may be used in next-generation MMM synthesis to replace conventional fillers. Each of these fillers shows attractive properties and attributes for future development.

In particular, the impregnation of carbonaceous nanofillers in polymer continuous matrix was deemed effective and very promising in the development of MMM for gas separation application. They have the ability to improve the engineering capability of the polymer matrix as well as effectively facilitate gas transport across the membrane. Yet, the full potential of these fillers has not been fully explored in gas separation membrane development, unlike water purification membrane technology. Despite the attractive physicochemical properties and gas separation performance of carbon nanotube (CNT) based MMM, CNT displayed poor dispersion characteristics in most polymer matrix domain and requires multiple functionalization or purification before incorporating into the polymer matrix. Instead of CNT, a viable alternative, namely carbon nanofiber (CNF) which exhibits similar properties as CNT, but improved dispersion degree is suggested. On the other hand, GO with excellent mechanical properties is found to exhibit good compatibility with most of the polymer host matrix. It could act as a selective barrier which impedes the direct diffusion of gases across the membrane and eventually enhances the gas separation efficiency of resultant MMM.

It is rather evident that incorporation technique employed for the development of POSS MMMs affects the membranes' gas permeability and selectivity. It was observed that most chemically crosslinked POSS MMMs exhibited increased permeability whilst its selectivity was compromised. The opposite is observed for the physically blended system. In a cross-linked system, the outer organic vertexes of POSS reacts with its neighboring polymer to form bonds. These organic vertexes are flexible and may cause a decrease in the crystallinity of the MMMs. On the other hand, in a physically blended POSS MMM, depending on the size of POSS aggregates formed, the crystallinity of the MMMs can either be enhanced or diminished. Larger aggregates of POSS act as an inhibitor to the segmental motion of the polymer matrix enhancing the glass transition temperature of MMMs. Hence, the decrease in MMMs crystallinity in a cross-linked POSS MMMs often lead to the increase in gas permeability. On the contrary, the increase in MMMs crystallinity in a physically blended POSS MMMs will lead to the decrement of gas permeability. Along with the importance of POSS dispersion, the performance of MMMs is also strongly dependent on the compatibility between POSS and polymer matrix. which dictates the morphology and crystallinity of the membrane. Thus, future studies should be focused on developing POSS MMMs which show simultaneously increased permeability and selectivity. To date, very few types of POSS has been utilized for the development of POSS MMMs in gas separation applications. Therefore, prospects on the development of newer and more efficient POSS MMMs using an array of POSS seem to be promising.

Meanwhile, the dispersion state of the clay platelets defines the morphology and gas separation performance of developed membranes. Techniques which are used to manufacture polymer composites can be applied to fabricate MMMs. For instance, melt compounding which may produce direct exfoliation without using organic solvents is an environmentally friendly, practical, and stable process that could be applied to fabricate MMMs dispersed with clay filler. Typically, the phase boundary defects between clay and polymer phase result in poor membrane performance. Nevertheless, these can be solved via crosslinking, thermal treatment and priming protocols. The orientation of layered silicates is also a critical factor in defining the gas separation performance of MMMs, thus it should be carefully controlled during membrane casting.

The incorporation of TiO_2 nanoparticles into polymer matrix has been found to improve MMM's gas separation performance. At optimal TiO_2 loading, the synthesized MMMs showed great improvement in gas permeance and selectivity as compared to those without the addition of TiO_2. Nonetheless, the existing problem with TiO_2 nanoparticles is their instability and low dispersibility due to strong surface interactions (Ng et al. 2013). Hence, the control of dispersion for TiO_2 nanoparticles in polymer matrix remains a challenge, especially at higher nanoparticles loading. Various organic functional groups can be further examined to be chemically linked onto TiO_2 particles in order to increase their surface stability and degree of dispersion in the polymer matrix.

Due to the inadequacy of single filler to enhance the membrane's overall separation performance as well as its tendency to agglomerate, a new approach of adding two fillers in the polymer matrix has emerged. Binary fillers' addition for MMM fabrication may improve filler dispersion without complex pre- or post-treatment, as well as enhance membrane gas separation performance in terms of permeability and selectivity. Therefore, future work may be expanded to develop binary filler-based MMM using fillers of different nature for gas separation application.

REFERENCES

Abedini, Reza, and Amir Nezhadmoghadam. 2010. "Application of membrane in gas separation processes: its suitability and mechanisms." *Petroleum & Coal* 52 (2):69-80.

Afzan Abdullah, Mohd, Hilmi Mukhtar, Hafiz Mannan, Yeong Yin Fong, and Maizatul Shima Shaharun. 2017. *Polyethersulfone/polyvinyl acetate blend membrane incorporated with TiO_2 nanoparticles for CO_2/CH_4 gas separation.* Vol. 13.

Ahmad, Jamil, Kalim Deshmukh, and May Britt Hägg. 2013. "Influence of TiO_2 on the Chemical, Mechanical, and Gas Separation Properties of Polyvinyl Alcohol-Titanium Dioxide (PVA-TiO_2) Nanocomposite Membranes." *International Journal of Polymer Analysis and Characterization* 18 (4):287-296. doi: 10.1080/1023666X.2013.7670 80.

Ahmadi, SJ, YD Huang, and W Li. 2004. "Synthetic routes, properties and future applications of polymer-layered silicate nanocomposites." *Journal of materials science* 39 (6):1919-1925.

Ahmadizadegan, Hashem. 2017. "Surface modification of TiO_2 nanoparticles with biodegradable nanocellulose and synthesis of novel polyimide/cellulose/TiO₂ membrane." *Journal of Colloid and Interface Science* 491:390-400. doi: 10.1016/j.jcis.2016.11.043.

Al-Saleh, Mohammed H, and Uttandaraman Sundararaj. 2009. "A review of vapor grown carbon nanofiber/polymer conductive composites." *Carbon* 47 (1):2-22.

Anadão, Priscila, Laís F Sato, Rafael R Montes, and Henrique S De Santis. 2014. "Polysulphone/montmorillonite nanocomposite membranes: Effect of clay addition and polysulphone molecular weight on the membrane properties." *Journal of Membrane Science* 455:187-199.

Annabi-Bergaya, Faïza. 2008. "Layered clay minerals. Basic research and innovative composite applications." *Microporous and Mesoporous Materials* 107 (1-2):141-148.

Aroon, MA, AF Ismail, T Matsuura, and MM Montazer-Rahmati. 2010. "Performance studies of mixed matrix membranes for gas separation: a review." *Separation and purification Technology* 75 (3):229-242.

Ayandele, Edunoluwa, Biswajit Sarkar, and Paschalis Alexandridis. 2012. "Polyhedral Oligomeric Silesqioxane (POSS)-Containing Polymer Nanocomposites" *Nanomaterials* 2:31. doi: 10.3390/nano2040445.

Azizi, Navid, Toraj Mohammadi, and Reza Mosayebi Behbahani. 2017. "Synthesis of a new nanocomposite membrane (PEBAX-1074/PEG-400/TiO₂) in order to separate CO₂ from CH₄." *Journal of Natural Gas Science and Engineering* 37:39-51. doi: https://doi.org/10.1016/j.jngse.2016.11.038.

Balandin, Alexander A, Suchismita Ghosh, Wenzhong Bao, Irene Calizo, Desalegne Teweldebrhan, Feng Miao, and Chun Ning Lau. 2008. "Superior thermal conductivity of single-layer graphene." *Nano letters* 8 (3):902-907.

Baney, Ronald H., Maki Itoh, Akihito Sakakibara, and Toshio Suzuki. 1995. "Silsesquioxanes." *Chemical Reviews* 95 (5):1409-1430. doi: 10.1021/cr00037a012.

Bastani, Dariush, Nazila Esmaeili, and Mahdieh Asadollahi. 2013. "Polymeric mixed matrix membranes containing zeolites as a filler for gas separation applications: A review." *Journal of Industrial and*

Engineering Chemistry 19 (2):375-393. doi: 10.1016/j.jiec.2012. 09.019.

Behroozi, Maryam, and Majid Pakizeh. 2017. "Study the effects of C loisite15 A nanoclay incorporation on the morphology and gas permeation properties of P ebax2533 polymer." *Journal of Applied Polymer Science* 134 (37):45302.

Bian, Yu, and Jovan Mijović. 2009. "Effect of side chain architecture on dielectric relaxation in polyhedral oligomeric silsesquioxane/ polypropylene oxide nanocomposites." *Polymer* 50 (6):1541-1547. doi: https://doi.org/10.1016/j.polymer.2009.01.036.

Bolotin, Kirill I, KJ Sikes, Zd Jiang, M Klima, G Fudenberg, J Hone, Ph Kim, and HL Stormer. 2008. "Ultrahigh electron mobility in suspended graphene." *Solid State Communications* 146 (9):351-355.

Brown, John F., and Lester H. Vogt. 1965. "The Polycondensation of Cyclohexylsilanetriol." *Journal of the American Chemical Society* 87 (19):4313-4317. doi: 10.1021/ja00947a016.

Bunch, J Scott, Scott S Verbridge, Jonathan S Alden, Arend M Van Der Zande, Jeevak M Parpia, Harold G Craighead, and Paul L Mceuen. 2008. "Impermeable Atomic Membranes from Graphene Sheets." *Nano Letters* 8 (8):5. doi: 10.1021/nl801457b.

Caro, Juergen, and Manfred Noack. 2008. "Zeolite membranes – Recent developments and progress." *Microporous and Mesoporous Materials* 115 (3):215-233. doi: 10.1016/j.micromeso.2008.03.008.

Chen-Yang, YW, YK Lee, YT Chen, and JC Wu. 2007. "High improvement in the properties of exfoliated PU/clay nanocomposites by the alternative swelling process." *Polymer* 48 (10):2969-2979.

Chen, Haibin, and David S Sholl. 2006. "Predictions of selectivity and flux for CH_4/H_2 separations using single walled carbon nanotubes as membranes." *Journal of Membrane Science* 269 (1):152-160.

Chung, Tai-Shung, Lan Ying Jiang, Yi Li, and Santi Kulprathipanja. 2007. "Mixed matrix membranes (MMMs) comprising organic polymers with dispersed inorganic fillers for gas separation." *Progress in Polymer Science* 32 (4):483-507. doi: 10.1016/j.progpolymsci.2007.01.008.

Cong, Hailin, Jianmin Zhang, Maciej Radosz, and Youqing Shen. 2007. "Carbon nanotube composite membranes of brominated poly (2, 6-diphenyl-1, 4-phenylene oxide) for gas separation." *Journal of Membrane Science* 294 (1):178-185.

Cordes, D. B., P. D. Lickiss, and F Rataboul. 2010. "Recent Developments in the Chemistry of Cubic Polyhedral Oligosilsesquioxane." *Chemical* 110:93.

Dai, Yan, Xuehua Ruan, Zhijun Yan, Kai Yang, Miao Yu, Hao Li, Wei Zhao, and Gaohong He. 2016. "Imidazole functionalized graphene oxide/PEBAX mixed matrix membranes for efficient CO_2 capture." *Separation and Purification Technology* 166:171-180.

De Azeredo, Henriette MC. 2009. "Nanocomposites for food packaging applications." *Food research international* 42 (9):1240-1253.

Defontaine, Guillaume, Anne Barichard, Sadok Letaief, Chaoyang Feng, Takeshi Matsuura, and Christian Detellier. 2010. "Nanoporous polymer–clay hybrid membranes for gas separation." *Journal of colloid and interface science* 343 (2):622-627.

Ebrahimi, Saeed, Shahram Mollaiy-Berneti, Hadi Asadi, Mohammad Peydayesh, Faranak Akhlaghian, and Toraj Mohammadi. 2016. "PVA/PES-amine-functional graphene oxide mixed matrix membranes for CO_2/CH_4 separation: Experimental and modeling." *Chemical Engineering Research and Design* 109:647-656.

Farrukh, Sarah, Sofia Javed, Arshad Hussain, and Muhammad Mujahid. 2014. "Blending of TiO_2 nanoparticles with cellulose acetate polymer: to study the effect on morphology and gas permeation of blended membranes." *Asia-Pacific Journal of Chemical Engineering* 9 (4):543-551. doi: doi:10.1002/apj.1783.

Feng, Lichao, Ning Xie, and Jing Zhong. 2014. "Carbon nanofibers and their composites: a review of synthesizing, properties and applications." *Materials* 7 (5):3919-3945.

Fu, Shao-Yun, Xi-Qiao Feng, Bernd Lauke, and Yiu-Wing Mai. 2008. "Effects of particle size, particle/matrix interface adhesion and particle loading on mechanical properties of particulate–polymer composites." *Composites Part B: Engineering* 39 (6):933-961.

Gain, O, E Espuche, E Pollet, M Alexandre, and Ph Dubois. 2005. "Gas barrier properties of poly (ε-caprolactone)/clay nanocomposites: Influence of the morphology and polymer/clay interactions." *Journal of Polymer Science Part B: Polymer Physics* 43 (2):205-214.

Galaleldin, S., H. A. Mannan, and H. Mukhtar. 2017. "Development and characterization of polyethersulfone/TiO_2 mixed matrix membranes for CO_2/CH_4 separation." *AIP Conference Proceedings* 1901 (1):130017. doi: 10.1063/1.5010577.

Galve, Alejandro, Daniel Sieffert, Claudia Staudt, Montserrat Ferrando, Carme Güell, Carlos Téllez, and Joaquín Coronas. 2013. "Combination of ordered mesoporous silica MCM-41 and layered titanosilicate JDF-L1 fillers for 6FDA-based copolyimide mixed matrix membranes." *Journal of Membrane Science* 431:163-170. doi: http://dx.doi.org/10. 1016/j.memsci.2012.12.046.

Ganesh, BM, Arun M Isloor, and Ahmad Fauzi Ismail. 2013. "Enhanced hydrophilicity and salt rejection study of graphene oxide-polysulfone mixed matrix membrane." *Desalination* 313:199-207.

Ge, Lei, Zhonghua Zhu, Feng Li, Shaomin Liu, Li Wang, Xuegang Tang, and Victor Rudolph. 2011. "Investigation of Gas Permeability in Carbon Nanotube (CNT)−Polymer Matrix Membranes via Modifying CNTs with Functional Groups/Metals and Controlling Modification Location." *The Journal of Physical Chemistry C* 115 (14):6661-6670. doi: 10.1021/jp1120965.

Ge, Lei, Zhonghua Zhu, and Victor Rudolph. 2011. "Enhanced gas permeability by fabricating functionalized multi-walled carbon nanotubes and polyethersulfone nanocomposite membrane." *Separation and Purification Technology* 78 (1):76-82. doi: 10.1016/j.seppur. 2011.01.024.

Gnanasekaran, D. 2016. "Nanocomposites of Polyhedral Oligomeric Silsesquioxane (POSS) and Their Application " In *Nanomaterials and Nanocomposites: Zero-to-Three Dimensional Material and Their Composites*, edited by P. M. Visakh and Maria J. M. Morlanes, 151-182. Weinheim, Germany: Wiley-VCH Verlag GmbH & Co. KGaA.

Goh, P. S., A. F. Ismail, S. M. Sanip, B. C. Ng, and M. Aziz. 2011a. "Recent advances of inorganic fillers in mixed matrix membrane for gas separation." *Separation and Purification Technology* 81 (3):243-264. doi: https://doi.org/10.1016/j.seppur.2011.07.042.

Goh, PS, AF Ismail, SM Sanip, BC Ng, and M Aziz. 2011b. "Recent advances of inorganic fillers in mixed matrix membrane for gas separation." *Separation and Purification Technology* 81 (3):243-264.

Hashemifard, SA, AF Ismail, and T Matsuura. 2011. "Effects of montmorillonite nano-clay fillers on PEI mixed matrix membrane for CO_2 removal." *Chemical Engineering Journal* 170 (1):316-325.

He, Qingliang, Lei Song, Yuan Hu, and Shun Zhou. 2009. "Synergistic effects of polyhedral oligomeric silsesquioxane (POSS) and oligomeric bisphenyl A bis(diphenyl phosphate) (BDP) on thermal and flame retardant properties of polycarbonate." *Journal of Materials Science* 44 (5):1308-1316. doi: 10.1007/s10853-009-3266-5.

Hegab, Hanaa M., and Linda Zou. 2015. "Graphene oxide-assisted membranes: Fabrication and potential applications in desalination and water purification." *Journal of Membrane Science* 484 (Supplement C):95-106. doi: https://doi.org/10.1016/j.memsci.2015.03.011.

Herring, Andrew M. 2006. "Inorganic–polymer composite membranes for proton exchange membrane fuel cells." *Journal of Macromolecular Science, Part C: Polymer Reviews* 46 (3):245-296.

Houshmand, Amirhossein, Wan Mohd Ashri Wan Daud, and Mohammad Saleh Shafeeyan. 2011. "Exploring potential methods for anchoring amine groups on the surface of activated carbon for CO_2 adsorption." *Separation Science and Technology* 46 (7):1098-1112.

Hu, Kesong, Dhaval D Kulkarni, Ikjun Choi, and Vladimir V Tsukruk. 2014. "Graphene-polymer nanocomposites for structural and functional applications." *Progress in Polymer Science* 39 (11):1934-1972.

Husain, Shabbir, and William J. Koros. 2007. "Mixed matrix hollow fiber membranes made with modified HSSZ-13 zeolite in polyetherimide polymer matrix for gas separation." *Journal of Membrane Science* 288 (1-2):195-207. doi: 10.1016/j.memsci.2006.11.016.

Ismail, AF, PS Goh, SM Sanip, and M Aziz. 2009a. "Transport and separation properties of carbon nanotube-mixed matrix membrane." *Separation and Purification Technology* 70 (1):12-26.

Ismail, Ahmad Fauzi, Pei Sean Goh, Suhaila Mohd Sanip, and Madzlan Aziz. 2009b. "Transport and separation properties of carbon nanotube-mixed matrix membrane." *Separation and Purification Technology* 70 (1):12-26.

Iyer, Pallavi, and Maria R. Coleman. 2008. "Thermal and mechanical properties of blended polyimide and amine-functionalized poly (orthosiloxane) composites." *Journal of Applied Polymer Science* 108 (4):2691-2699. doi: 10.1002/app.27607.

Jaafar, Juhana, AF Ismail, T Matsuura, and K Nagai. 2011. "Performance of SPEEK based polymer–nanoclay inorganic membrane for DMFC." *Journal of membrane science* 382 (1-2):202-211.

Jadav, Ghanshyam L., and Puyam S. Singh. 2009. "Synthesis of novel silica-polyamide nanocomposite membrane with enhanced properties." *Journal of Membrane Science* 328 (1-2):257-267. doi: 10.1016/j. memsci.2008.12.014.

Jamil, Asif, Oh Pei Ching, and Azmi Shariff. 2014. "Polymer-Nanoclay Mixed Matrix Membranes for CO_2/CH_4 Separation: A Review." *Applied Mechanics & Materials* (625).

Jamil, Asif, Oh Pei Ching, and Azmi BM Shariff. 2016. "Current Status and Future Prospect of Polymer-Layered Silicate Mixed-Matrix Membranes for CO_2/CH_4 Separation." *Chemical Engineering & Technology* 39 (8):1393-1405.

Jamil, Asif, Oh Pei Ching, and Azmi M Shariff. 2017. "Mixed matrix hollow fibre membrane comprising polyetherimide and modified montmorillonite with improved filler dispersion and CO_2/CH_4 separation performance." *Applied Clay Science* 143:115-124.

Jamil, Asif, Oh Pei Ching, and Azmi M Shariff. 2018. "Polyetherimide-montmorillonite mixed matrix hollow fibre membranes: Effect of inorganic/organic montmorillonite on the CO_2/CH_4 separation." *Separation and Purification Technology*.

Ji, Xuqiang, Yuanhong Xu, Wenling Zhang, Liang Cui, and Jingquan Liu. 2016. "Review of functionalization, structure and properties of graphene/polymer composite fibers." *Composites: Part A* 87:17. doi: 10.1016/j.compositesa.2016.04.011.

Johnson, David W, Ben P Dobson, and Karl S Coleman. 2015. "A manufacturing perspective on graphene dispersions." *Current Opinion in Colloid & Interface Science* 20 (5):367-382.

Julian, Helen, and IG Wenten. 2012. "Polysulfone membranes for CO_2/CH_4 separation: State of the art." *IOSR Journal of Engineering* 2 (3):484-495.

Khabashesku, Valery N. 2011. "Covalent functionalization of carbon nanotubes: synthesis, properties and applications of fluorinated derivatives." *Russian Chemical Reviews* 80 (8):705-725.

Kiadehi, A. Dehghani, M. Jahanshahi, A. Rahimpour, and A. A. Ghoreyshi. 2014. "Fabrication and Evaluation of Functionalized Nano-titanium Dioxide (F-NanoTiO$_2$)/ polysulfone (PSf) Nanocomposite Membranes for Gas Separation." *Iranian Journal of Chemical Engineering(IJChE)* 11 (4):40-49.

Kim, Jaemyung, Laura J Cote, Franklin Kim, Wa Yuan, Kenneth R Shull, and Jiaxing Huang. 2010. "Graphene oxide sheets at interfaces." *Journal of the American Chemical Society* 132 (23):8180-8186.

Kim, Sangil, Liang Chen, J Karl Johnson, and Eva Marand. 2007. "Polysulfone and functionalized carbon nanotube mixed matrix membranes for gas separation: theory and experiment." *Journal of Membrane Science* 294 (1):147-158.

Kim, Sangil, Todd W Pechar, and Eva Marand. 2006. "Poly (imide siloxane) and carbon nanotube mixed matrix membranes for gas separation." *Desalination* 192 (1-3):330-339.

Kim, Wun-gwi, Sunho Choi, and Sankar Nair. 2011. "Swelling, functionalization, and structural changes of the nanoporous layered silicates AMH-3 and MCM-22." *Langmuir* 27 (12):7892-7901.

Kim, Wun-gwi, Jong Suk Lee, David G Bucknall, W. J. Koros, and Sankar Nair. 2013. "Nanoporous Layered Silicate AMH-3/Cellulose Acetate

Nanocomposite Membranes for Gas Separation." *Journal of Membrane Science* 441:8. doi: 10.1016/j.memsci.2013.03.044.

Kingston, Christopher T., Zygmunt J. Jakubek, Stéphane Dénommée, and Benoit Simard. 2004. "Efficient laser synthesis of single-walled carbon nanotubes through laser heating of the condensing vaporization plume." *Carbon* 42 (8):1657-1664. doi: https://doi.org/10.1016/j.carbon. 2004.02.020.

Kojima, Yoshitsugu, Arimitsu Usuki, Masaya Kawasumi, Akane Okada, Toshio Kurauchi, Osami Kamigaito, and Keisuke Kaji. 1995. "Novel preferred orientation in injection-molded nylon 6-clay hybrid." *Journal of Polymer Science Part B: Polymer Physics* 33 (7):1039-1045.

Konios, Dimitrios, Minas M Stylianakis, Emmanuel Stratakis, and Emmanuel Kymakis. 2014. "Dispersion behaviour of graphene oxide and reduced graphene oxide." *Journal of colloid and interface science* 430:108-112.

Konnertz, Nora, Yi Ding, Wayne J. Harrison, Peter M. Budd, Andreas Schönhals, and Martin Böhning. 2017. "Molecular mobility and gas transport properties of nanocomposites based on PIM-1 and polyhedral oligomeric phenethyl-silsesquioxanes (POSS)." *Journal of Membrane Science* 529:274-285. doi: 10.1016/j.memsci.2017.02.007.

Kuo, Shiao-Wei, and Feng-Chih Chang. 2011. "POSS related polymer nanocomposites." *Progress in Polymer Science* 36 (12):1649-1696. doi: 10.1016/j.progpolymsci.2011.05.002.

Li, Guizhi, Lichang Wang, Hanli Ni, and Charles U. Pittman. 2001. "Polyhedral Oligomeric Silsesquioxane (POSS) Polymers and Copolymers: A Review." *Journal of Inorganic and Organometallic Polymers* 11 (3):123-154. doi: 10.1023/a:1015287910502.

Li, Jiang, Matthew J Vergne, Eric D Mowles, Wei-Hong Zhong, David M Hercules, and Charles M Lukehart. 2005. "Surface functionalization and characterization of graphitic carbon nanofibers (GCNFs)." *Carbon* 43 (14):2883-2893.

Li, Jing-Feng, Zhen-Liang Xu, Hu Yang, Li-Yun Yu, and Min Liu. 2009. "Effect of TiO_2 nanoparticles on the surface morphology and performance of microporous PES membrane." *Applied Surface Science* 255 (9):4725-4732. doi: https://doi.org/10.1016/j.apsusc.2008.07.139.

Li, Xueqin, Youdong Cheng, Haiyang Zhang, Shaofei Wang, Zhongyi Jiang, Ruili Guo, and Hong Wu. 2015. "Efficient CO_2 capture by functionalized graphene oxide nanosheets as fillers to fabricate multi-permselective mixed matrix membranes." *ACS applied materials & interfaces* 7 (9):5528-5537.

Li, Xueqin, Lu Ma, Haiyang Zhang, Shaofei Wang, Zhongyi Jiang, Ruili Guo, Hong Wu, XingZhong Cao, Jing Yang, and Baoyi Wang. 2015a. "Synergistic effect of combining carbon nanotubes and graphene oxide in mixed matrix membranes for efficient CO_2 separation." *Journal of Membrane Science* 479:1-10.

Li, Xueqin, Lu Ma, Haiyang Zhang, Shaofei Wang, Zhongyi Jiang, Ruili Guo, Hong Wu, XingZhong Cao, Jing Yang, and Baoyi Wang. 2015b. "Synergistic effect of combining carbon nanotubes and graphene oxide in mixed matrix membranes for efficient CO_2 separation." *Journal of Membrane Science* 479:1-10.

Li, Yi, and Tai-Shung Chung. 2010. "Molecular-level mixed matrix membranes comprising Pebax® and POSS for hydrogen purification via preferential CO_2 removal." *International Journal of Hydrogen Energy* 35 (19):10560-10568. doi: https://doi.org/10.1016/j.ijhydene.2010.07.124.

Li, Yuanqing, Tianyi Yang, Ting Yu, Lianxi Zheng, and Kin Liao. 2011. "Synergistic effect of hybrid carbon nanotube-graphene oxide as a nanofiller in enhancing the mechanical properties of PVA composites." *Journal of Materials Chemistry* 21 (29):10844-10851.

Liang, Chia-Yu, Petr Uchytil, Roman Petrychkovych, Yung-Chieh Lai, Karel Friess, Milan Sipek, M. Mohan Reddy, and Shing-Yi Suen. 2012. "A comparison on gas separation between PES (polyethersulfone)/MMT (Na-montmorillonite) and PES/TiO_2 mixed matrix membranes." *Separation and Purification Technology* 92:57-63. doi: 10.1016/j.seppu r.2012.03.016.

Liu, Chun Qing, S. Kulprathipanja, Alexis M. W. Hillock, and W. J. Koros. 2008. "Membrane Materials and Characterisation." In *Advance*

Membrane Technology and Applications, edited by T. Matsuura, W.S Winston Ho, Anthony G Fane and Norman N Li, 788-819. John Wiley & Sons, Inc.

Lozano, K, J Bonilla-Rios, and EV Barrera. 2001. "A study on nanofiber-reinforced thermoplastic composites (II): Investigation of the mixing rheology and conduction properties." *Journal of Applied Polymer Science* 80 (8):1162-1172.

Madhavan, K., and B. S. R. Reddy. 2009. "Structure–gas transport property relationships of poly(dimethylsiloxane–urethane) nanocomposite membranes." *Journal of Membrane Science* 342 (1-2):291-299. doi: 10.1016/j.memsci.2009.07.002.

Mahmoud, Khaled A., Bilal Mansoor, Ali Mansour, and Marwan Khraisheh. 2015. "Functional graphene nanosheets: The next generation membranes for water desalination." *Desalination* 356 (Supplement C):208-225. doi: https://doi.org/10.1016/j.desal.2014.10.022.

Mansur, NA. 2014. "*Effects of Aluminosilicate Mineral Nano-Clay Fillers on Polysulfone Mixed Matrix Membrane for Carbon Dioxide Removal.*"

Matteucci, Scott, Victor A. Kusuma, David Sanders, Steve Swinnea, and Benny D. Freeman. 2008. "Gas transport in TiO_2 nanoparticle-filled poly(1-trimethylsilyl-1-propyne)." *Journal of Membrane Science* 307 (2):196-217. doi: 10.1016/j.memsci.2007.09.035.

Milliman, Henry W., Hatsuo Ishida, and David A. Schiraldi. 2012. "Structure Property Relationships and the Role of Processing in the Reinforcement of Nylon 6-POSS Blends." *Macromolecules* 45 (11):4650-4657. doi: 10.1021/ma3002214.

Mittal, Garima, Vivek Dhand, Kyong Yop Rhee, Soo-Jin Park, and Wi Ro Lee. 2015. "A review on carbon nanotubes and graphene as fillers in reinforced polymer nanocomposites." *Journal of Industrial and Engineering Chemistry* 21:11-25. doi: http://dx.doi.org/10.1016/j.jiec.2014.03.022.

Mittal, Vikas. 2009. "Polymer layered silicate nanocomposites: a review." *Materials* 2 (3):992-1057.

Moghadam, F., M. R. Omidkhah, E. Vasheghani-Farahani, M. Z. Pedram, and F. Dorosti. 2011. "The effect of TiO_2 nanoparticles on gas transport properties of Matrimid5218-based mixed matrix membranes." *Separation and Purification Technology* 77 (1):128-136. doi: https://doi. org/10.1016/j.seppur.2010.11.032.

Moradihamedani, Pourya, Nor Azowa Ibrahim, Wan Md Zin Wan Yunus, and Nor Azah Yusof. 2015. "Study of morphology and gas separation properties of polysulfone/titanium dioxide mixed matrix membranes." *Polymer Engineering & Science* 55 (2):367-374. doi: 10.1002/pen. 23887.

Nasir, R., H. Mukhtar, Z. Man, and D. F. Mohshim. 2013. "Material Advancements in Fabrication of Mixed-Matrix Membranes." *Chemical Engineering & Technology* 36 (5):717-727. doi: 10.1002/ceat.2012 00734.

Ng, Law Yong, Abdul Wahab Mohammad, Choe Peng Leo, and Nidal Hilal. 2013. "Polymeric membranes incorporated with metal/metal oxide nanoparticles: A comprehensive review." *Desalination* 308:15-33. doi: https://doi.org/10.1016/j.desal.2010.11.033.

Nguyen, Quang T, and Donald G Baird. 2006. "Preparation of polymer–clay nanocomposites and their properties." *Advances in Polymer Technology: Journal of the Polymer Processing Institute* 25 (4):270-285.

Okada, Akane, and Arimitsu Usuki. 1995. "The chemistry of polymer-clay hybrids." *Materials Science and Engineering: C* 3 (2):109-115.

Park, Sungjin, and Rodney S Ruoff. 2009. "Chemical methods for the production of graphenes." *Nature nanotechnology* 4 (4):217-224.

Pavlidou, S, and CD Papaspyrides. 2008. "A review on polymer-layered silicate nanocomposites." *Progress in polymer science* 33 (12):1119-1198.

Pei, Songfeng, and Hui-Ming Cheng. 2012. "The reduction of graphene oxide." *Carbon* 50 (9):3210-3228.

Pervin, Farhana, Yuanxin Zhou, Vijaya K Rangari, and Shaik Jeelani. 2005. "Testing and evaluation on the thermal and mechanical properties of carbon nano fiber reinforced SC-15 epoxy." *Materials Science and Engineering: A* 405 (1):246-253.

Phillips, Shawn H., Timothy S. Haddad, and Sandra J. Tomczak. 2004. "Developments in nanoscience: polyhedral oligomeric silsesquioxane (POSS)-polymers." *Current Opinion in Solid State and Materials Science* 8 (1):21-29. doi: 10.1016/j.cossms.2004.03.002.

Picard, Emilie, H Gauthier, J-F Gérard, and Eliane Espuche. 2007. "Influence of the intercalated cations on the surface energy of montmorillonites: consequences for the morphology and gas barrier properties of polyethylene/montmorillonites nanocomposites." *Journal of Colloid and Interface Science* 307 (2):364-376.

Qiu, Shi, Liguang Wu, Guozhong Shi, Lin Zhang, Huanlin Chen, and Congjie Gao. 2010. "Preparation and pervaporation property of chitosan membrane with functionalized multiwalled carbon nanotubes." *Industrial & Engineering Chemistry Research* 49 (22):11667-11675.

Quan, Shuai, Song Wei Li, You Chang Xiao, and Lu Shao. 2017. "CO_2-selective mixed matrix membranes (MMMs) containing graphene oxide (GO) for enhancing sustainable CO_2 capture." *International Journal of Greenhouse Gas Control* 56:22-29.

Raftopoulos, Konstantinos N., and Krzysztof Pielichowski. 2016. "Segmental dynamics in hybrid polymer/POSS nanomaterials." *Progress in Polymer Science* 52:136-187. doi: 10.1016/j.progpolymsci.2015.01.003.

Rezaei, M, AF Ismail, SA Hashemifard, and T Matsuura. 2014. "Preparation and characterization of PVDF-montmorillonite mixed matrix hollow fiber membrane for gas-liquid contacting process." *Chemical Engineering Research and Design* 92 (11):2449-2460.

Rezakazemi, Mashallah, Abtin Ebadi Amooghin, Mohammad Mehdi Montazer-Rahmati, Ahmad Fauzi Ismail, and Takeshi Matsuura. 2014a. "State-of-the-art membrane based CO_2 separation using mixed matrix membranes (MMMs): an overview on current status and future directions." *Progress in Polymer Science* 39 (5):817-861.

Rezakazemi, Mashallah, Abtin Ebadi Amooghin, Mohammad Mehdi Montazer-Rahmati, Ahmad Fauzi Ismail, and Takeshi Matsuura. 2014b. "State-of-the-art membrane based CO_2 separation using mixed matrix membranes (MMMs): an overview on current status and future directions." *Progress in Polymer Science* 39 (5):817-861.

Ríos-Dominguez, H., F. A. Ruiz-Treviño, R. Contreras-Reyes, and A. González-Montiel. 2006. "Syntheses and evaluation of gas transport properties in polystyrene–POSS membranes." *Journal of Membrane Science* 271 (1-2):94-100. doi: 10.1016/j.memsci.2005.07.014.

Sadeghi, Morteza, Hajar Taheri Afarani, and Zohreh Tarashi. 2015. "Preparation and investigation of the gas separation properties of polyurethane-TiO$_2$ nanocomposite membranes." *Korean Journal of Chemical Engineering* 32 (1):97-103. doi: 10.1007/s11814-014-0198-9.

Sadeghi, Morteza, and Hajar Taheri. 2014. *Preparation and investigation of the gas separation properties of polyurethane-TiO$_2$ nanocomposite membranes.* Vol. 32.

Safaei, P., A. Marjani, and M. Salimi. 2016. "Mixed matrix membranes prepared from high impact polystyrene with dispersed TiO$_2$ nanoparticles for gas separation." *Journal of Nanostructures* 6 (1):74-79. doi: 10.7508/jns.2016.01.012.

Sánchez-Soto, M., David A. Schiraldi, and S. Illescas. 2009. "Study of the morphology and properties of melt-mixed polycarbonate–POSS nanocomposites." *European Polymer Journal* 45 (2):341-352. doi: 10.1016/j.eurpolymj.2008.10.026.

Sánchez, M., J. Rams, M. Campo, A. Jiménez-Suárez, and A. Ureña. 2011. "Characterization of carbon nanofiber/epoxy nanocomposites by the nanoindentation technique." *Composites Part B: Engineering* 42 (4):638-644. doi: http://dx.doi.org/10.1016/j.compositesb.2011.02.017.

Sarfraz, Muhammad, and M Ba-Shammakh. 2016a. "Combined effect of CNTs with ZIF-302 into polysulfone to fabricate MMMs for enhanced CO$_2$ separation from flue gases." *Arabian Journal for Science and Engineering* 41 (7):2573-2582.

Sarfraz, Muhammad, and M Ba-Shammakh. 2016b. "Synergistic effect of adding graphene oxide and ZIF-301 to polysulfone to develop high performance mixed matrix membranes for selective carbon dioxide separation from post combustion flue gas." *Journal of Membrane Science* 514:35-43.

Scott, Donald W. 1946. "Thermal Rearrangement of Branched-Chain Methylpolysiloxanes1." *Journal of the American Chemical Society* 68 (3):356-358. doi: 10.1021/ja01207a003.

Shamsabadi, Ahmad Arabi, Farzad Seidi, Ehsan Salehi, Mohammad Nozari, Ahmad Rahimpour, and Masoud Soroush. 2017. "Efficient CO_2-removal using novel mixed-matrix membranes with modified TiO_2 nanoparticles." *Journal of Materials Chemistry A* 5 (8):4011-4025. doi: 10.1039/C6TA09990D.

Shen, Jie, Gongping Liu, Kang Huang, Wanqin Jin, Kueir-Rarn Lee, and Nanping Xu. 2015. "Membranes with fast and selective gas-transport channels of laminar graphene oxide for efficient CO_2 capture." *Angewandte Chemie* 127 (2):588-592.

Silvestre, Clara, Donatella Duraccio, and Sossio Cimmino. 2011. "Food packaging based on polymer nanomaterials." *Progress in polymer science* 36 (12):1766-1782.

Skoulidas, Anastasios I, David S Sholl, and J Karl Johnson. 2006. "Adsorption and diffusion of carbon dioxide and nitrogen through single-walled carbon nanotube membranes." *The Journal of chemical physics* 124 (5):054708.

Striolo, Alberto, Clare McCabe, and Peter T. Cummings. 2005. "Thermodynamic and Transport Properties of Polyhedral Oligomeric Silsesquioxanes in Poly(dimethylsiloxane)." *The Journal of Physical Chemistry B* 109 (30):14300-14307. doi: 10.1021/jp045388p.

Sung, Jun Hee, Hyun Suk Kim, Hyoung-Joon Jin, Hyoung Jin Choi, and In-Joo Chin. 2004. "Nanofibrous membranes prepared by multiwalled carbon nanotube/poly (methyl methacrylate) composites." *Macromolecules* 37 (26):9899-9902.

Takahashi, Nobuyuki, and Kazuyuki Kuroda. 2011. "Materials design of layered silicates through covalent modification of interlayer surfaces." *Journal of Materials Chemistry* 21 (38):14336-14353.

Tanaka, Kazuo, Shigehiro Adachi, and Yoshiki Chujo. 2009. "Structure-property relationship of octa-substituted POSS in thermal and mechanical reinforcements of conventional polymers." *Journal of*

Polymer Science Part A: Polymer Chemistry 47 (21):5690-5697. doi: 10.1002/pola.23612.

Tanaka, Kazuyoshi, and Sumio Iijima. 2014. *Carbon nanotubes and graphene*: Newnes.

Tewari, P. K. 2016. *Nanocomposite Membrane Technology, Fundamentals and Applications*. New York: Taylor & Francis Group.

Tibbetts, Gary G, Ioana C Finegan, and Choongyoong Kwag. 2002. "Mechanical and electrical properties of vapor-grown carbon fiber thermoplastic composites." *Molecular Crystals and Liquid Crystals* 387 (1):129-133.

Tibbetts, Gary G, Max L Lake, Karla L Strong, and Brian P Rice. 2007. "A review of the fabrication and properties of vapor-grown carbon nanofiber/polymer composites." *Composites Science and Technology* 67 (7):1709-1718.

Tjong, Sie Chin. 2012. *Polymer composites with carbonaceous nanofillers: properties and applications*: John Wiley & Sons.

Tran, Phong A, Lijie Zhang, and Thomas J Webster. 2009. "Carbon nanofibers and carbon nanotubes in regenerative medicine." *Advanced drug delivery reviews* 61 (12):1097-1114.

Tsetseris, L., and S. T. Pantelides. 2014. "Graphene: An impermeable or selectively permeable membrane for atomic species?" *Carbon* 67:58-63. doi: 10.1016/j.carbon.2013.09.055.

Uddin, Faheem. 2008. "Clays, nanoclays, and montmorillonite minerals." *Metallurgical and Materials Transactions A* 39 (12):2804-2814.

Usuki, Arimitsu, Akihiko Koiwai, Yoshitsugu Kojima, Masaya Kawasumi, Akane Okada, Toshio Kurauchi, and Osami Kamigaito. 1995. "Interaction of nylon 6-clay surface and mechanical properties of nylon 6-clay hybrid." *Journal of Applied Polymer Science* 55 (1):119-123.

Valero, Marta, Beatriz Zornoza, Carlos Téllez, and Joaquín Coronas. 2014. "Mixed matrix membranes for gas separation by combination of silica MCM-41 and MOF NH 2-MIL-53 (Al) in glassy polymers." *Microporous and Mesoporous Materials* 192:23-28.

Van Noorden, Richard. 2006. *Moving towards a graphene world*. Nature Publishing Group.

Vatanpour, Vahid, Sayed Siavash Madaeni, Ali Reza Khataee, Ehsan Salehi, Sirus Zinadini, and Hossein Ahmadi Monfared. 2012. "TiO_2 embedded mixed matrix PES nanocomposite membranes: Influence of different sizes and types of nanoparticles on antifouling and performance." *Desalination* 292:19-29. doi: https://doi.org/10.1016/j.desal.2012.02. 006.

Villaluenga, JPG, M Khayet, MA Lopez-Manchado, JL Valentin, B Seoane, and JI Mengual. 2007. "Gas transport properties of polypropylene/clay composite membranes." *European Polymer Journal* 43 (4):1132-1143.

Wang, Ting, Cai-hong Yang, Chun-Li Man, Li-guang Wu, Wan-Lei Xue, Jiang-nan Shen, Bart Van der Bruggen, and Zhuan Yi. 2017. "Enhanced Separation Performance for CO_2 Gas of Mixed-Matrix Membranes Incorporated with TiO_2/Graphene Oxide: Synergistic Effect of Graphene Oxide and Small TiO_2 Particles on Gas Permeability of Membranes." *Industrial & Engineering Chemistry Research* 56 (31):8981-8990.

Weng, Tzu-Hsiang, Hui-Hsin Tseng, and Ming-Yen Wey. 2009. "Preparation and characterization of multi-walled carbon nanotube/ PBNPI nanocomposite membrane for H_2/CH_4 separation." *international journal of hydrogen energy* 34 (20):8707-8715.

Xin, Qingping, Zhao Li, Congdi Li, Shaofei Wang, Zhongyi Jiang, Hong Wu, Yuan Zhang, Jing Yang, and Xingzhong Cao. 2015. "Enhancing the CO_2 separation performance of composite membranes by the incorporation of amino acid-functionalized graphene oxide." *Journal of Materials Chemistry A* 3 (12):6629-6641.

Xue, Qingzhong, Xinglong Pan, Xiaofang Li, Jianqiang Zhang, and Qikai Guo. 2017. "Effective enhancement of gas separation performance in mixed matrix membranes using core/shell structured multi-walled carbon nanotube/graphene oxide nanoribbons." *Nanotechnology* 28 (6):065702.

Yang, Leixin, Zhizhang Tian, Xiyuan Zhang, Xingyu Wu, Yingzhen Wu, Yanan Wang, Dongdong Peng, Shaofei Wang, Hong Wu, and Zhongyi Jiang. 2017. "Enhanced CO_2 selectivities by incorporating CO_2-philic PEG-POSS into polymers of intrinsic microporosity membrane."

Journal of Membrane Science 543:69-78. doi: 10.1016/j.memsci.2017. 08.050.

Yu, Yunwu, Wenhao Pan, Xiaoman Guo, Lili Gao, Yaxin Gu, and Yunxue Liu. 2017. "A poly(arylene ether sulfone) hybrid membrane using titanium dioxide nanoparticles as the filler: Preparation, characterization and gas separation study." *High Performance Polymers* 29 (1):26-35. doi: 10.1177/0954008315626990.

Zhang, Xianfeng, Bin Wang, Jaka Sunarso, Shaomin Liu, and Linjie Zhi. 2012. "Graphene nanostructures toward clean energy technology applications." *Wiley interdisciplinary reviews: energy and environment* 1 (3):317-336.

Zhang, Yuan, Jaka Sunarso, Shaomin Liu, and Rong Wang. 2013. "Current status and development of membranes for CO_2/CH_4 separation: A review." *International Journal of Greenhouse Gas Control* 12:84-107.

Zhao, Dan, Jizhong Ren, Ying Wang, Yongtao Qiu, Hui Li, Kaisheng Hua, Xinxue Li, Jiemei Ji, and Maicun Deng. 2017. "High CO_2 separation performance of Pebax®/CNTs/GTA mixed matrix membranes." *Journal of Membrane Science* 521:104-113. doi: 10.1016/j.memsci. 2016.08.061.

Zinadini, Sirus, Ali Akbar Zinatizadeh, Masoud Rahimi, Vahid Vatanpour, and Hadis Zangeneh. 2014. "Preparation of a novel antifouling mixed matrix PES membrane by embedding graphene oxide nanoplates." *Journal of Membrane Science* 453:292-301.

Zornoza, Beatriz, Beatriz Seoane, Juan M Zamaro, Carlos Téllez, and Joaquín Coronas. 2011. "Combination of MOFs and zeolites for mixed-matrix membranes." *ChemPhysChem* 12 (15):2781-2785.

Zulhairun, A. K., and A. F. Ismail. 2014a. "The role of layered silicate loadings and their dispersion states on the gas separation performance of mixed matrix membrane." *Journal of Membrane Science* 468:20-30. doi: 10.1016/j.memsci.2014.05.038.

Zulhairun, AK, and AF Ismail. 2014b. "The role of layered silicate loadings and their dispersion states on the gas separation performance of mixed matrix membrane." *Journal of Membrane Science* 468:20-30.

Zulhairun, AK, AF Ismail, T Matsuura, MS Abdullah, and A Mustafa. 2014. "Asymmetric mixed matrix membrane incorporating organically modified clay particle for gas separation." *Chemical Engineering Journal* 241:495-503.

In: Gas Separation
Editor: Suraya Mathews

ISBN: 978-1-53614-606-6
© 2019 Nova Science Publishers, Inc.

Chapter 3

METHANE PURIFICATION FROM LANDFILL USING PVC BASED MEMBRANE

Sunarti Abd Rahman, Raj Krishna Roshan, Norhidayana Mandayar, Nadia Sofea Hazleen and Sureena Abdullah

Faculty of Chemical & Natural Resources Engineering,
Universiti Malaysia Pahang, 26300 Gambang, Pahang, Malaysia

ABSTRACT

Mixed-Matrix Membrane is a developing technology that has been use in the gas separation process due to the ability of MMMs to cope with the limitation of polymeric membrane and inorganic membrane. Therefore, this research is conducting to study the permeability and selectivity of carbon dioxide (CO_2) and methane gas (CH_4) of polyvinylchloride (PVC) Mixed-Matrix-Membrane (MMMs) with the inorganic fillers of zeolite 4Å particles. The fabrication of MMMs is prepared by using dry/wet phase inversion method. Fourier Transform Infrared Spectroscopy (FTIR) is used to study the chemical interaction of the membrane by analyzing the intensity of the peak of chloride vibration. Meanwhile, Scanning Electron Microscope (SEM) is use to analyses the cross sectional morphology of MMMs. The performance of MMMs analyses by using Design of Expert

(DOE) method. While, the model regression equation is developed as the potential use for screening the permeability of CO_2 and CH_4 based on the effect of PVC and zeolite concentration.

Keywords: mixed-matrix membrane, selectivity of CO_2/CH_4, permeability of CO_2, permeability of CH_4

INTRODUCTION

Typically, landfill gas or biogas derived from a landfill comprises generally equally molal amounts of a mixture of carbon dioxide (CO_2) and methane (CH_4), with the CO_2 and CH_4 representing about 90 mole percent or more of the biogas. The landfill gas also contains minor amounts of nitrogen, oxygen, hydrogen, carbon monoxide and a variety of undesirable trade impurities present at the ppm level, as well as water vapour. The nitrogen and oxygen content of the biogas depends on the air ingress to the landfill- and gas-collection system. Elimination of contaminant CO_2 from natural gas and landfill gas streams, composed mostly of CH_4, is an important problem. The presence of CO_2 in natural gas significantly lowers the energy density of the gas stream and can lead to pipeline corrosion over time [1].

Biogas, specifically landfill gas, often contains too much CO_2 and too low a CH_4 concentration to fuel a natural gas engine for electrical power generation. CH4 gas stands as one of the most prevalent gaseous in the air. Global CH_4 emissions from landfill are estimated to be between 30 and 70 million tonnes each year [2]. CH_4 originating from landfill is vastly found in developed and populated countries, where the levels of waste tend to be the highest. CH_4's unique role as a greenhouse gas and as the primary component of natural gas means that reducing CH_4 emissions can yield significant economic, environmental and operational benefits. Companies are reducing their emissions of greenhouse gases, improving operational safety and enhancing the efficiency of their operations. Further economic and operational benefits can result when CH_4 mitigation activities reduce maintenance and fuel requirements or result in the capture of other valuable hydrocarbon resources. Currently in the market 1 m^3 of CH$_4$ is equivalent to $68.93 [3]. A recent study was carried out at Kampung Sg. Ikan Landfill in Kuala Terengganu, Terengganu, Malaysia. Municipal solid wastes from the city of Kuala Terengganu were gathered, weighed and datas were collected. Wastes were segregated according to their 13 types, such as 3D plastic, 2D plastic, glass, and so on (Table 1). CH$_4$ can be achieved or obtained through solid waste, mainly food waste, through the precise processes and through gas separation.

Among the many types of polymers that exhibit gas separation properties for gas mixtures, a few polymers such as polysulfone (PSf), polyethersulfone (PESf), polyetherimide (PEI) and polyimide (PI) have been recognized as promising polymers with respect to their permeability and selectivity. Nowadays, polymeric membranes dominate the industry because of the outstanding economy and competitive performance [7]. The membranes can be operated at ambient temperature and they have good mechanical and chemical properties [8]. PVC is a polymer with a wide variety of applications in different industries due to high compatibility with additives, easy process ability and recyclability. Despite the extensive use of PVC in several industrial applications, studies on the gas separation performance of PVC membranes are scarce [9]. Despite their suitability for various applications in research and commercialization, polymeric membranes are still ineffective in meeting the requirement for the current advanced membrane technology as these materials have demonstrated a trade-off between the permeability and selectivity, with an 'upper-bound' evident as proposed by Robeson [10].

To overcome the limitation in of polymeric membrane and inorganic membrane in gas separation, mixed-matrix membranes (MMMs) is develop as an alternative approach in gas separation process. MMMs are a hybrid membrane of organic-inorganic compounds that proposing better separation performance at reasonable price. The combination of both membranes gives a pleasant stability of molecular sieving and better performance of organic membrane [11]. The MMMs characterized by embedding inorganic materials into the polymeric matrix which can be any polymeric materials such as polysulfone, poly-vinyl-hloride and polyamides. Since MMMs are design to improve the host membrane of polymeric membrane, the selection of inorganic material is important due to the advantages of peculiar properties of inorganic fillers such as zeolite [12]. Therefore, the main factor in the fabrication of MMMs is the affinity between the two phases involved and the compatibility of hybrid organic-inorganic membrane.

Over the past decade, researcher use zeolite as an inorganic filler in the fabrication of MMMs because the characteristics of zeolite in having well-defined size, uniform pore distribution, high specific area and high prorsity [11]. Nonetheless, in phase separation, the critical issue found in the development of MMMs is the poor compatibility and adhesion of zeolite-polymer. In other hand, the fabrication of MMMs with zeolite based is highly cost since the preparation on the modification of zeolite is difficult in large-scale production. Therefore, due to the issues, a study is conducting to develop a newly combination of zeolite- PVC in the fabrication of MMMS.

Table 1. Municipal solid wastes data from Kg. Sg. Ikan landfill for 3 weeks

Categories/Week	1	2	3	Total
Total waste (kg)	11010.00	13500.00	7040.00	31550.00
Waste weighed (kg)	9926.95	12453.20	7007.90	29388.05
Waste weighed (%)	90.16	92.25	99.54	93.15
Plastic 3D (kg)	362.59	480.00	274.30	1116.89
Plastic 2D (kg)	869.10	1164.50	665.30	2698.90
Metal (kg)	7.96	33.70	33.20	74.86
Aluminum can (kg)	100.61	195.90	107.30	403.81
Paper (kg)	843.60	878.40	741.30	2463.30
Pampers (kg)	873.35	1631.00	889.40	3393.75
Glass (kg)	200.72	235.70	142.10	578.52
Wood/Landscape (kg)	1099.51	2179.20	1802.30	5081.01
Polystrene (kg)	118.90	219.70	81.10	419.70
Bed (kg)	9.48	0.00	10.00	19.48
Textile (kg)	108.94	289.50	146.40	544.84
Food waste (kg)	4542.91	5227.60	2107.60	11878.11
Others (kg)	1063.62	1046.80	66.50	2176.92

METHODS

Experimental

Chemicals

Polyvinyl chloride (PVC) is a polymeric material which grouped as a glassy polymer that will be use in the fabrication of membrane, which have been purchased from Sigma Aldrich. The properties of PVC in having high operating temperature enable the gas separation process work at maximum condition. N-Methyl-2-pyrrolidone (NMP) is use as a solvent in sample preparation, NMP which has chemical formula of C_5H_9NO and the average molecular weight of 99.13g/mol with 99.5% [13]. NMP is purchased from Sigma Aldrich. The molecular sieve involved in the preparation of Mixed-Matrix-Membranes (MMMs) is zeolite 4Å. Type 4Å is the sodium form of molecular sieve which it will absorb those molecules having a critical diameter of less than 4Å. The molecular sieve has particle size of 8 - 12 mesh purchased from Aldrich. Methane gas and Carbon dioxide gas with purity 99% were obtained from Air Products Malaysia Sdn. Bhd for gas permeation.

Sample Preparation

The approach method for the preparation of mixed-matrix membrane is through dry/wet phase inversion. The zeolite is prepared from 5 to 10 W/V% and dissolve it in varied volume of NMP with 90cm^3 to 95cm^3. Then, the mixture will mix homogeneously by stirring the mixture for three hours at condition of 200 rpm and 100°C. Weight the PVC with 5 to 10 W/V% and add the PVC to the solution until it reaches homogeneous mixture. In order to achieve yield homogeneous mixture, agitate the mixture at 300 rpm by using magnetic stirrer at the room temperature for another three hours. Store the dope solution at storage vessel and degasses all the bubble by leaving the dope solution overnight at room temperature. Prepare glass plate for the casting method by using glass rod. Finally, the sample membrane was ready for the permeation test for calculate the permeability and selectivity of CO_2 and CH_4 separation.

Single Gas Permeation Test for CO$_2$/CH$_4$ Separation

In this study the membrane use for each Single Gas Permeation Test has a diameter approximately 6cm. First, purified CH$_4$ at two bars was used as a gas test and connect the membrane with the gas permeation apparatus manufactured by Aba Manufacturing Sdn Bhd. Then, set the temperature at room temperature (25°C ± 5°C) and finally, measure the gas permeation rate by using soap-bubble meter manufactured by Dwyer. Repeat the procedures of Single Gas Permeation Test by replacing the test gas with CO$_2$ and set the pressure at one bar. Solubility of the pure gas is a factor that will affect the rate permeation gas. Therefore, CO$_2$ will be the last gas to be measure due to the encouragement of membrane plasticization.

Screening Study of Membrane for CO$_2$/CH$_4$ Separation

The experimental data is screening by using Design of Experiment (DOE) software, version 7.1. DOE is software that enables the users to interpret multi-factor experiments. This software promotes a wide range of design including factorials, fractional factorials and composite design. Meanwhile, in analysing an experiment DOE fit the model that relating with the response or quality characteristics to a set of controllable variables. In this study there are three factors were selected which are concentration of zeolite (X1), concentration of PVC (X2) and duration time of immersion for glass plate in water (X3) and selectivity of CO$_2$ and CH$_4$ as response which coded as Y in the software. The selectivity of CO$_2$ and CH$_4$ that diffuse through the membrane analyse as response which denote as Y in the software. Each response parameter that containing notable term will be develop in mathematical model by using multiple linear regression analysis (MLRA) and analysis of variance (ANOVA).

RESULT AND DISCUSSION

Gas Performance Analysis

Single gas permeation test is used to study the permeability and selectivity of CO$_2$ and CH$_4$ for all samples of fabricated membrane. Three samples from each different eight composition of formulated membrane with vary concentration of zeolite, PVC and NMP were tested to get the average reading of result. The complete experiment for each of the sample and output response were tabulated at Table 2. The proportion differentiation between the whole design matrixes is less than 5% for all of the three samples of membrane. The range of the selectivity of CO_2/CH_4 is between 2.2777 to 2.8377. The condition of the single gas permeation test is 1bar. At low operating pressure the solubility coefficient of CO_2 was greater which promoted more permeation rate of CO_2 over polymeric dense membrane. Therefore, at 1bar single gas permeation test is most favourable method to test the permeability and selectivity of CO_2 and CH_4.

Table 2. Experimental design matrix and response results

Specimen	Concentration Variable			Experimental result
	X_1 (W/V %)	X_2 (W/V %)	Time, t (min)	Y
1	5	5	5	2.6039
2	10	5	5	2.6638
3	5	10	5	2.9539
4	10	10	5	2.6254
5	5	5	15	2.4281
6	10	5	15	2.6247
7	5	10	15	2.2777
8	10	10	15	2.8377

X_1, Concentration of zeolite; X_2, Concentration of PVC; and duration time of immersion for glass plate in water and Y, selectivity of CO_2/CH_4

Fourier Transform Infrared Spectroscopy (FTIR) Analysis

Fourier Transform Infrared Spectroscopy (FTIR) spectra is using to detect the presence of chemical added to the MMMs and the interactions between PVC and zeolite particles [14, 15, 16]. Figure 1 show characteristics for the sample of fabricated membrane with the highest loading of zeolite (10% W/V%) correspond with 5 W/V% of PVC of 15 min duration time for immersion of glass plate due the performance of the fabricated membrane obeys the upper bound Robeson plot. The selectivity of CO_2/CH_4 of fabricated membrane is 2.6247 with respect to the significant morphology for polymerization of membrane. From Figure 1, the peak value 3640 cm^{-1} is corresponding to the O-H bound in zeolite. The interaction shows that the bound between polymer and zeolite exists in the sample of fabricated membrane. However, the elimination of void spaces between the polymer matrix and zeolite are not enough through this interaction. Aromatic carbon double bond, C=O bare associated with the peak number is 1563cm^{-1} which indicates the aromatic bonds were not disrupted and remained in the membrane. Meanwhile, the band located at 2877 cm^{-1} was associated with CH_3 stretch vibration of the membrane. Meanwhile, the band located at 2877 cm^{-1} anhydride C-O stretch in the PVC/Zeolite mixed-matric membrane.

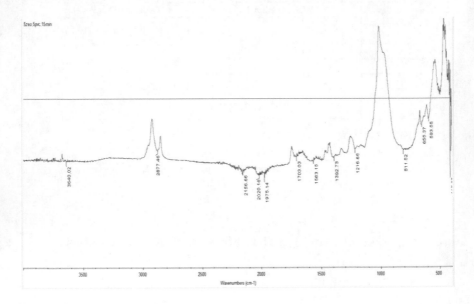

Figure 1. FTIR spectra of PVC/Zeolite mixed-matrix membrane.

Table 3 demonstrated the summary for the characteristics of adsorption for the sample with the selectivity of CO_2/CH_4.

Table 3. Functional group of membrane sample with selectivity of CO_2/CH_4 in respect to the comparison of characteristics adsorption

Functional Group	Sample	Characteristics Adsorption
CH_3 stretch	2877.46	2865-2885
Aromatic carbon double bond C=O	1563.15	1600 or 1475
Anhydride C-O stretch	1216.66	1300
O-H bound in zeolite	3640	3200-3600

Scanning Electron Microscope (SEM)

SEM is used to study the morphology of the cross-sectional area of fabricated membrane. Figure 2 shows the cross-sectional area SEM image of PVC/Zeolite membrane taken from sample for 15 min duration time of immersion rate of 10 W/V% zeolite, 5 W/V% PVC. The Figure 2(a) show the SEM image of PVC/Zeolite MMMs with 2000X magnified while Figure 2(b) is more clear SEM image for the morphology of zeolite molecules. The presence zeolite molecules as a filler in polymeric membrane increases the adhesion of the polymer chain. Zeolite is an inorganic filler which enhance the flexibility of the polymeric membrane by improving the interaction between the heterogeneous phases of the matrix membrane. Furthermore, the adhesion of organic-inorganic phase interaction is improved with the presence of thin layer around of chemical composition the zeolite molecules as a chemical treatment for the matrix membrane.

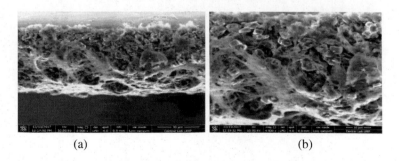

(a)   (b)

Figure 2. SEM images of the Mixed-matrix membrane of Zeolite/PVC; (a) 2000x magnified SEM image of PVC/Zeolite MMMs (b) 400x magnified SEM image of PVC/Zeolite MMMs.

ANOVA Analysis

The manipulated data of the study is expressed with the variable of concentration for Zeolite (X1), amount of Polyvinylchloride (X2), and time taken for immersion of membrane (X3) with correspond output of selectivity of CO_2/CH_4 (Y3). This research approach two statistical point of view to analysed and asses the model of the experimental data by using significance of factor (SOF) and R-squared (R^2) test show the result for SOF and the interaction for the selectivity of CO_2/CH_4 with the value of R^2 for the analysed model is $R^2 > 0.7$ which is the variation of the response could be justified by using mathematical modelling. From Table 4, the original value of R^2, adjusted R^2 and predicted R^2 are demonstrated after neglecting the insignificance terms of the design model. The predicted R^2 is 0.948237 is in reasonable agreement with adjusted R^2 because the value is within 0.2. In other words, the design model of the experiment is in an accurate description of experimental data which indicated the relationship between the variables and response data. The model result shows the tendencies for the model to form linear regression fit which showed that the experimental research range is adequate.

Table 4. R^2 statistic for the fitted model

Model source	Selectivity of CO_2/CH_4
Std. Dev.	0.023085
Mean	2.530733
Coeddicient of variation	0.912191
R-Squared	0.996765
Adj R-Squared	0.988677
Pred R-Squared	0.948237
Adeq Precision	31.53732

Empirical Model Analysis

The approach model is well fitted to the case study (experimental result), since from Table 4 the value of $R^2 > 0.90$ which is 0.988677. In other words, the selectivity of CO_2/CH_4 is reliable to the regression model of the membrane's permeability. Meanwhile, Table 5 demonstrated the result of F-test ANOVA from the regression model with 95% confidence level. The model value of F-value is 123.241 and the prob > F is less than 0.0500 which implies that the model is significant. In other words, the improvement of the experimental model is achieved by discarding insignificant effects term after completing all the eights sample. Table 5 show the result of F-test ANOVA for the selectivity of CO_2/CH_4.

Table 5. ANOVA statistic for the fitted model

	Sum of Squares	DOF	Mean Square	F- Value	p-value Prob > F
Model	0.32839	5	0.06568	123.241	0.0081
A-X1: Concentration of Zeolite	0.19759	1	0.19759	370.776	0.0027
B-X2: Concentration of PVC	0.0195	1	0.0195	36.5852	0.0263
AB	0.06929	1	0.06929	130.0181	0.0076
AC	0.008182	1	0.008182	15.35316	0.0594
BC	0.033827	1	0.033827	63.47446	0.0154
Residual	0.001066	2	0.000533		

At 10% Zeolite loading (highest loading of zeolite) the void size is greater than the diameter of the gas molecules which enable the gas molecules to pass through (penetrate) the voids with lower diffusion resistance instead of selective pores of zeolite.

The interaction of A: concentration of zeolite and B: concentration of PVC (AB) have the most significant effect on the selectivity of CO_2/CH_4 with the F-value is 130.0181 and the P-value is only 0.0076% which is less than 0.05%. The interaction of zeolite as an inorganic membrane with PVC (polymeric membrane) is significant because of adhesion of hybrid membrane. During the casting process, the PVC is detached from the zeolite surface producing the micro-cavities throughout the membrane. However, functionalized molecular sieves are required to achieve high selectivity of CO_2/CH_4 and good performance of mixed-matrix membrane due to the great interaction between the polymeric membrane phase and sieves.

Meanwhile, the interaction of AC and BC where C: time duration for immersion of membrane in minute resulted is significant with the f-value are 15.3516 and 63.4766 respectively. While the p-value for AC interaction is 0.0594 which is slightly from 0.05% for p-value and BC interaction is 0.0154 which less than 0.05% to achieve significant data.

Verification on Statistical Models and Diagnostic Statistic

The interaction between independent factors in the experimental model is being investigated through Response surface method to observe the effects among the variables.

Figure 3 demonstrated the interaction between the independent variables by combining two independent factors in binary combination for all the responses. Figure 3(a) show that the interaction between A: Concentration of zeolite and B: Concentration on of PVC with the diagonal data obtained shows that as the increment of AB interaction increase the selectivity of CO_2/CH_4. Meanwhile, Figure 3(b) the interaction of AC demonstrated the same pattern of AB interaction where C: duration of immersion time of glass plate in minute. Meanwhile, the standard deviation of the experimental model is test through demonstrated plotted data of the normal probability of the residual as shown in Figure 4(a). The plotting data is important to ensure the actual and predicted response value is still obeying the normal distribution. The residual data versus predicted is plotted in Figure 4(b) with the data scattered randomly in constant range of residual across the graph within the horizontal line. It shows that the experimental model proposed is feasible and constant amount of variance is confirmed.

In other hands, the reliability of the empirical data is confirmed from the comparison of predicted output data from regression model with the actual values obtained from experimental results as shown in Figure 4(c). Generally, the predicted value is directly proportional to the actual value of the experimental results. In other words, the value of the predicted data increases as the increment of the selectivity of CO_2/CH_4 as well as the actual value of the experimental results. The output result from the experimental response is well fitted in acceptable variance range. Therefore, the regression model obtained from DOE is enabling to further as a predictor for the analysis of variability of membrane concentration to produce high selectivity of CO_2/CH_4.

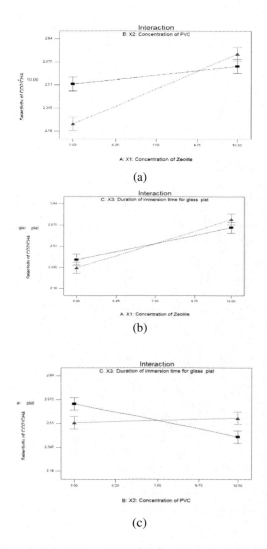

(a)

(b)

(c)

Figure 3. Interaction via Selectivity of CO/CH for (a) concentration of PVC and zeolite (b) concentration of Zeolite and Duration of immersion time for glass plat (c) Concentration of PVC and Duration of immersion time for glass plat.

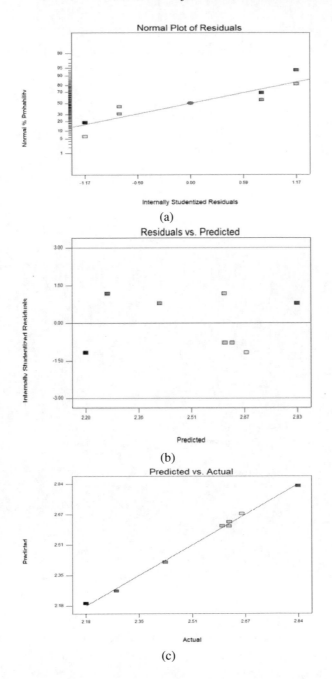

Figure 4. (a) Normal probability plot of residual; (b) Plot of residual versus predicted response and (c) Predicted vs. actual values plot.

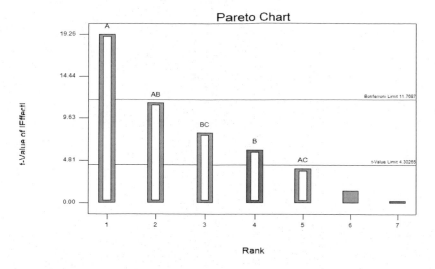

Figure 5. Pareto chart.
A: Concentration of Zeolite (W/V%)
B: Concentration of PVC (W/V%)
C: duration time of immersion for glass plate in water.

From Figure 5, the Pareto chart shows the f-value from the ANOVA which demonstrated that there is a variation on the value of effects on the difference type of factors of the model. The Pareto chart show that the factor of A is 19.26 which is the highest amongst the others and it at the first rank in Pareto chart. Meanwhile, the interaction between AB, BC and AC show the different f-value which are 11.7687, 1.64 and 4.30265 respectively. The factor B is insignificant in the Pareto chart as it exists in blue colour which ANOVA analysis indicated it as insignificant factor. The Bonferron limit is at 11. 7687 which the peak of AB interaction. While, the t-Value limit is at the peak of the interaction of AC.

Model Equations Based on Screening Effect

Figure 3 demonstrated the interaction for all the factors in the experimental model which showed that the membrane is performed in linear model. The factors of the model are in significant model term as well as the selectivity of CO_2/CH_4. The model term is A; Concentration of Zeolite (W/V%), B; Concentration of PVC (W/V%) an and C; Time taken for immersion of membrane in minute. From Table 5, the F-test ANOVA analysis performed the regression to assess the equation of the experimental model with 95% confidence level.

Equation 1 expressed the regression model equation and coefficient for selectivity of CO_2/CH_4.

$$\text{Selectivity of } CO_2/CH_4 = 2.53 + 0.16A - 0.049B + 0.093AB + 0.032AC + 0.065BC \qquad (1)$$

where;

A: Concentration of zeolite

B: Concentration of PVC

C: Duration time of immersion for glass plate in water

CONCLUSION

The fabrication of Poly-Vinyl-chloride and zeolite are to investigate the effect of polymer concentration and additive of (zeolite) on the permeability of CO_2 and CH_4 as well as the selectivity of CO_2/CH_4. The homogeneous mixture of polymer and zeolite were miscible with each other. From FTIR result the existence of 3640 cm^{-1} peak explained the presence of of an interaction between polymer and zeolite. The highest selectivity of CO_2/CH_4 are observed to be 2.8377. meanwhile, the result from DOE expressed that these mixed-matrix membranes are affected with the interaction of concentration of zeolite, concentration of PVC and duration time for the immersion rate of glass plate to offer high selectivity of CO_2/CH_4 by using ANOVA analysis. The regression equation is developed from regression model is expressed in [Equation 1 for the selectivity of CO_2/CH_4 based on the effect of polymer concentration.

ACKNOWLEDGMENTS

The authors wish to thank Universiti Malaysia Pahang for the grant (RDU 1803113), Faculty of Chemical and Natural Resources Engineering for the Gas Engineering lab facilities.

REFERENCES

[1]   Omar, K. Farha., Youn-Sang, Bae., Brad, G. Hauser., Alexander, M. Spokoyny., Randall, Q. Snurr., Chad, A. Mirkin. & Joseph, T. Hupp. (2010). "Chemical reduction of a diimide based porous polymer for selective uptake of carbon dioxide versus methane". *Chemical Community*, *46*, 1056-1058. DOI:10.1039/B922554D.

[2]   Chen, Xiao Yuan., Serge, Kaliaguine. & Denis, Rodrigue. (2017). "Correlation between Performances of Hollow Fibers and Flat Membranes for Gas Separation". *Separation & Purification Reviews*, *47.1*, 66-87. DOI:10.1080/15422119.2017.1324490.

[3]   Stefan, Lechtenböhmer., Carmen, Dienst., Manfred, Fischedick., Thomas, Hanke., Roger, Fernandez., Don, Robinson., Ravi, Kantamaneni. & Brian, Gillis. (2009). "Tapping the leakages: Methane losses, mitigation options and policy issues for Russian long-distance gas transmission pipelines". *International Journal of Greenhouse Gas Control*, *1.4*, 387-395. DOI:10.1016/S1750-5836 (07)00089-8.

[4]   Richard, W. Baker. & Bee, Ting Low. (2014). "Gas separation membrane materials: a perspective". *Macromolecules*, *47*, 6999-7013. DOI: 10.1021/ma501488s.

[5]   Suresh, K. Bhargava., Sharifah, Bee Hamid. & Sridhar, S. (2014). *"Membrane-Based Gas Separation: Principle, Applications, and Future Potentials"*. https://pdfs.semanticscholar.org/3829/9fb9007 22280197fedbccbbd5d8682b405ac.pdf.

[6]   Angelo, Basile. & Francesco, Gallucci. (2011). *"Introduction to Membranes, Membranes for Membrane Reactors: Preparation,*

Optimization and Selection". Chichester: John Wiley and Sons Inc, 45-46. DOI: 10.1002/9780470977569.

[7] Daniel, R. Dreyer., Christopher, W. Bielawski. & Alexander, D. Todd. (2010). "The chemistry of graphene oxide". *Chemical Society Review*, *39*, 228-240. DOI:10.1039/C4CS00060A.

[8] Luis, M. Gandía., Gurutze, Arzamendi. & Pedro, M. Diéguez. (2013). "Renewable Hydrogen Technologies: Production, Purification, Storage, Applications and Safety". *Newnes*, 156-157, 211. https://doi.org/10.1002/ente.201402151.

[9] Mohammad, Mohagheghian., Morteza, Sadeghi., Mahdi, Pourafshari. & Chenar, Mahdi Naghsh. (2014). "Gas separation properties of polyvinylchloride (PVC)-silica nanocomposite membrane". *Korean Journal of Chemical Engineering*, *31.11*, 2041-2050. DOI: 10.1007/s11814-014-0169-1.

[10] Lloyd, M. Robeson. (1991). "Correlation of separation factor versus permeability for polymeric membranes". *Journal of Membrane Science*, *62*, 165. DOI: 10.1016/0376-7388(91)80060-J.

[11] Norwahyu Jusoh, Y. F. (2016). "Enhanced gas separation performance using mixed matrix membranes". *Journal of Membrane Science*, *525*, 175-186. DOI: 10.1016/j.memsci.2016.10.044.

[12] Gloria, M. Monsalve-Bravo. & Suresh, K. Bhatia. (2017). "Extending effective medium theory to finite size systems: Theory and simulation for permeation in mixed-matrix membranes". *Journal of Membrane Science*, 148-149. DOI: 10.1016/j.memsci.2017.02. 029.

[13] Carretti, E. e. (2013). "Synthesis and characterization of gels from polyallylamine and carbon dioxide as Gellant". *J. Am. Chem. Soc*, 5121-5129. DOI: 10.1021/ja034399d.

[14] Chen, Chee Lek. & Sunarti, Abd Rahman. (2015). "Formulation of Mixed-Matrix Membrane (PSF/Zeolite) for CO_2/N_2 Separation". *Journal of Materials Science and Chemical Engineering*, 67-68. DOI: 10.4236/msce.2015.35008.

[15] Abtin, Ebadi Amooghin, Hamidreza, Sanaeepur., Mona, Zamani Pedram Mohammadreza. & Ali, Kargari. (2016). "New advances in polymeric membranes for CO_2 separation". *Polymer science: research*

advances, practical applications and educational aspects (A. Méndez-Vilas; A. Solano, Eds.), 354. Chapter: www.formatex.info/polymerscience1/book/354-368.pdf.

[16] Fawziea, M. Hussein., Dr. Amel, S. Merzah. & Zaid, W. Rashad. (2014). "Preparation of PVC Hollow fiber membrane using (DMAC/Acetone)". *Journal of Chemical and Petroleum Engineering, 87.* Available online at: www.iasj.net.

BIOGRAPHICAL SKETCH

Sunarti Abd Rahman

Affiliation: Faculty of Chemical & Natural Resources Engineering, Universiti Malaysia Pahang, 26300 Gambang, Pahang, Malaysia.
Education: Sunarti, PhD
Research and Professional Experience: Membrane Technology, Gas Separation Technology, Innovative Material, Waste water Treatment
Professional Appointments: Senior Lecturer

Publications from the Last 3 Years:

[1] Chen, Chee Lek. & **Sunarti**, **Abd Rahman**. (2015). Formulation of Mixed-Matrix Membrane (PSF/Zeolite) for CO2/N2 Separation: Screening of Polymer Concentration. *Journal of Materials Science and Chemical Engineering, 3,* 65-74.

[2] **Sunarti**, **Abdul Rahman**. & Wan Zulaisa, Amira Wan Jusoh. (2015). Optimization of the Preparation of Hydrophobic Isotactic Polypropylene Flat Sheet Membrane by Response Surface Methodology Design. *Journal of Scientific Research & Reports,* 7-12.

[3] Wan Zulaisa, Amira Wan Jusoh. & **Sunarti**, **Abdul Rahman**. (2015). Preparation Isotactic polyropylene Hydrophobic Microporous Flat Sheet via Tips for Membrane Contactor. *Advanced Material Research, 1113,* 36-42, ISSN: 1662- 8958.

[4] Khalid, T. Rashid., **Sunarti**, **Abdul Rahman**. & Qusay, F. Alsalhy. (2015). Hydrophobicity Enhancement Of Poly (Vinylidene Fluoride-*co-* Hexafluoro Propylene) for Membrane Distillation. *Journal of Polymer Science and Technology*, 1, 1-9.

[5] Abdul Wahab, M. S. & **Sunarti**, **A. R**. (2015). Production of Mixed Matrix (PVDF/Zeolite) Membrane for CO2/N2 Gas Separation. *International Journal of Chemical and Biomolecular Science*, 1, 264-270.

[6] Abdul Wahab, M. S. & **Sunarti**, **A. R**. (2015). Development of PEBAX Based Membrane for Gas Separation: A Review. *International Journal of Membrane Science and Technology*, 2, 78-84.

[7] Khalid, T. Rashid. & **Sunarti**, **Binti Abdul Rahman**. (2016). Enhancement the flux of PVDF-co-HFP hollow fiber Membranes for direct contact membrane distillation applications. *ARPN Journal of Engineering and Applied Sciences*, 4 (11), 2189 – 2192.

[8] Khalid, T. Rashid., **Sunarti**, **Abdul Rahman**. & Qusay, F. Alsalhy. (2016). Optimum Operating Parameters for Hollow Fiber Membranes in Direct Contact Membrane Distillation. *Arabian Journal for Science & Engineering*, 7(41), 2647- 2658.

[9] Wan Zulaisa, Amira Wan Jusoh., **Sunarti**, **Abdul Rahman**. & Rosmawati, Naim. (2016). The effects of adipic acid on the hydrophobicity IPP membranes prepared using DPE via TIPS. *ARPN Journal of Engineering and Applied Sciences*, 10 (11), 6376 – 6383.

[10] Abdul Wahab, M. S., **Sunarti**, **A. R**. & Nurul Farhana, D. (2016). Preliminary investigation on gas separation ability of polysulfone/ pebax 1657 composite membrane. *Jurnal Teknologi*, 78(11), 155-160.

[11] Abdul Wahab, M. S. & **Sunarti**, **A. R**. (2017). Influence of PVDF/Pebax TFC Casting Temperature towards CO2/N2 Gas Separation. *Indian Journal of Science and Technology*, 10, 2, 20-25.

[12] Mohamad, Syafiq Abdul Wahab., **Sunarti**, **Abdul Rahman**. & Abdul, Latif Ahmad. (2017). Biomethane Purification Using PVDF/Pebax 1657 Thin Film Composite Membrane. *Journal of Physical Science*, *28*, 1, 39–51.

[13] Mohamad, Syafiq Abdul Wahab. & **Sunarti**, **Abd Rahman**. (2018). The Effect Number of Pebax 1657 Coating Layer on Thin Film Composite (TFC) Membrane for CO2/N2 Separation. *Chiang Mai J. Sci.*, 2018, *45*(1), 484-491.

INDEX

R

S

T